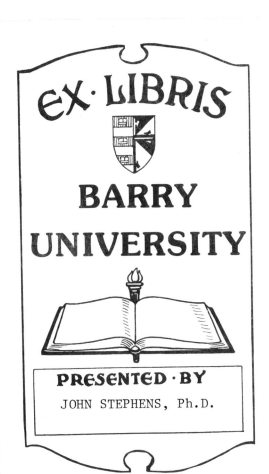

Computers and the Representation
of Geographical Data

Computers and the Representation of Geographical Data

E. E. Shiryaev

All-Union Hydrotechnical and Meliorative Research Institute Moscow

Translated from the Russian by

V. M. Divid, N. N. Protsenko, and Yu. U. Rajabov

A Wiley–Interscience Publication

JOHN WILEY & SONS

Chichester . New York . Brisbane . Toronto . Singapore

Originally published in Russian
© 1977 by Nedra Publishers, Moscow
This edition copyright © 1987 by John Wiley & Sons Ltd.

Library of Congress Cataloging-in-Publication Data:

Shiryaev, E. E. (Evgeniĭ Evgen'evich)
    Computers and the representation of geographical data.
    Translation of: Kartograficheskoe otobrazhenie
preobrazovanie geoinformatŝii.
    'A Wiley–Interscience publication'.
    Bibliography: p.
    1. Cartography—Data processing.   I. Title.
GA102.4.E4S5513 1987   526'.028'5   86-13342
ISBN 0 471 90915 7

British Library Cataloguing in Publication Data:

Shiryaev, E. E.
    Computers and the representation of
    geographical data.
    1. Cartography—Data processing.
    I. Title
    526'.028'54   GA102.4.E4
ISBN 0 471 90915 7

Printed and bound in Great Britain

191820

Contents

Introduction

Cartography as a tool for investigating natural phenomena has in recent years acquired great importance. This has been facilitated not only by an increased interest in the study of natural resources, associated with space exploration, but primarily by the progress of cartography itself in mathematizing and automating the representation, storage, retrieval, and analysis of data concerning the Earth.

This monograph is devoted to the questions of elaborating new methods in cartographic representation of geographical data, taking into account the image-perceiving capacity not only of man but also of computers, and developing computer techniques to plot, read, and analyse maps.

Examined in the monograph are the methods used to achieve complete automation of the process of geographical data representation and analysis, starting from a computer-assisted reading of the initial map and covering the plotting of a derived map, constructing a digital model, and obtaining numerical values as a result of cartographic data analysis.

Main emphasis is placed on methods of raising the efficiency of the use of maps with the help of modern automatic devices. Particular attention has been given to simplifying the map plotting process, providing a reliable computerized identification of symbols by means of the new cartographic representation methods, with the traditional visual clarity being preserved.

When a map becomes intelligible, not only to man but also to the computer, extensive possibilities open up for its automatic analysis, i.e. for investigating natural and socioeconomic phenomena using modern cybernetic facilities. With the customary graphical form of geographical data representation being preserved, a map will function as a common universal language mutually understood by man and computer (in this case, data transfer onto intermediate information carriers, such as magnetic tape or disc, is unnecessary). Despite the progress in the development of digital models, the cartographic form of data representation remains the basic one.

Vast information flows arriving via aerospace survey channels impose more stringent demands on the promptitude of data interpretation and the plotting of maps. This can only be achieved with the help of digital methods of automatic data processing. The efficiency of aerospace data interpretation can be raised by utilizing the presently available cartographic materials. Joint processing of aerospace photographs together with large-scale maps considerably facilitates photography control, interpretation, and creation of digital data banks. Accordingly, the book pays special attention, along with designing new types of maps, to using traditional maps, primarily for constructing digital models. Different techniques of preparing traditional maps for computerized development of digital models, based on the principles of image normalization, are also reviewed.

Considering the similarity of the principles underlying automatic processing of aerospace photographs and maps, their analysis and interpretation by different methods are described, among them the techniques of transforming the shape of objects (mainly extended ones) to be represented in the form of derivative maps. The principles governing the automatic processing of images on aerospace photographs are partially dealt with. Possibilities of combined use of different methods for interpreting telesounding and cartographic data in solving the problems of evaluating the state and predicting the future of complex natural processes are shown, as exemplified by land reclamation and water management problems.

The book is primarily concerned with the results of the author's own research, representing a new scientific trend in cartography and aimed at creating a unified common cartographic language for man and computer. Among these are the method of raster discretization, implicit optical coding of images for automatic readout, general aspects of the theory of cartographic modelling and generalization, techniques of multiscale representation of areal and linear objects, generalization of areal (combined techniques) and extended objects, methods of automatic transformation of images, automatic calculation of metric parameters from maps, and a number of other methods. The basic principles of this approach were for the first time systematically presented in the author's book [120].

Since then different types of normalized maps have been designed and published in small numbers. A software complex for their automatic compiling and reading by a computer (conversion into a digital model) has been developed. Practical implementation of the above techniques shows that in their outward appearance these maps can be the same as the traditional ones, but they can simultaneously carry encoded information easily perceived by computer. While many of the representation techniques are basically meant for automatic processing, they proved to be more effective for detailed and accurate representation of data, without impairing the visual clarity and readability of images. Combinations of traditional and normalized representation techniques, aimed at raising the visual clarity of images, are examined.

The author is very pleased that this monograph is to be published in the country where the Oxford Automated Cartographic System—the first such system in the world—was created. He would like to express his hope that the new trend in cartography, presented in the book, will be of interest to English-speaking readers.

CHAPTER 1
General Aspects of the Theory of Cartographic Representation, Transformation, and Analysis of Geographical Data

By the nature of the problems solved and the information processed cartographic modelling procedures can be subdivided into solving the direct problem—cartographic representation of geographical data (the designing, compiling and printing of maps) and the inverse problem—utilization of cartographic information (reading, interpretation, etc.).

The need to develop means and methods for the utilization of cartographic information at the present-day level of science and technology is now felt very sharply. The effectiveness of the practical application of cartographic information can be regarded as the main stimulating factor for designing maps and developing cartography as a whole. This determines its usefulness and ability to compete with modern non-cartographic methods of computer analysis of geographical data presented in digital form.

A significant role in the development of cartography is played by the methods of geographical data representation (MGDR). MGDR to a certain extent determine, on the one hand, the detail, accuracy and authenticity of geographical data and, on the other hand, the possibility of automating their processing. Efficient utilization of geographical information carried by a map depends on how suitable its form is for automatic readout and processing on a computer. In this connection the problem of developing MGDR should be regarded as integrated with designing computer hardware components and identification algorithms.

1.1. Possibilities for Improving a Map as a Geographical Information Carrier

Is a map up-to-date as a geographical information carrier? Does it meet the requirements of our time—the time of global investigation of our planet, the Earth, and the other planets of the Solar system, when space vehicles are used to obtain initial information about them, and high-speed computers—to process it? No, it is not! The traditional map is an inconvenient carrier of information to be transmitted to, and analysed by, a computer.

For a man the traditional cartographic form of representing information is customary and easy to read, whereas for a computer it is very complicated. This is quite understandable, since the traditional map has always been designed to be read by man only, and so naturally it proved to be barely suitable for automatic reading and analysis. That is why, in a number of sciences studying the Earth, it is preferable to record the results of field surveys from the very beginning in a digital form on computer-oriented media with subsequent automatic analysis, i.e. omitting the cartographic form of data representation. A map, if plotted in this case, performs illustrative functions only.

This situation continues despite the fact that cartographic methods are used today in conjunction with modern automatic and computer facilities and aerospace survey methods. To solve successfully such important problems as prompt representation, analysis and forecasting of the state of objects at a certain moment of time, and the dynamics of their development it is essential to automate not only the process of representing objects or phenomena on maps (where cartography has in fact made its main achievements), but also their readout and analysis. Cartographic representation and analysis of dynamic processes such as soil erosion, pollution of land waters, changes in vegetation cover under the action of harmful man-made or natural factors, etc., especially need the use of automation facilities interacting closely with tele-sounding methods. This demand becomes particularly obvious if one bears in mind that cartographic methods of data analysis make it possible, using a combination of maps differing in their content, to characterize, generalize, and compare the totality of positive and negative processes and phenomena. At the same time, in solving this problem, regardless of the application of even the most modern facilities for cartographic data analysis, it is not always possible to obtain results with sufficient accuracy and promptness, which is usually the case with the traditional form of data representation. Hence there naturally arises the necessity of searching for new pathways in the theory and practice of data representation that would provide the means not only for visual analysis, but also for reliable automatic analysis, of maps. In reaching this objective the most complicated problem is that of the automatic reading of maps.

The basic difficulty lies in the identification of symbols [25, 41, 57], designed on traditional maps only for visual perception without any allowance for the resolving power and the logical facilities of computers. Man's capabilities in searching for and identifying images are known to be much greater than those of a computer. And although today we can read automatically and with reasonable reliability certain types of traditional maps where information is presented by a comparatively limited number of symbols of a simple pattern, the imagery on a number of maps is so complicated (particularly when multicoloured symbols of intricate pattern intersect or come into contact) that it is impossible as yet for them to be reliably identified by a computer. Because of the difficulty and unreliability of identifying symbols, and due to long computer times, the transformation of cartographic data in digital form to be input into and recorded by a computer is performed manually with the help of encoders (digitizers). It is unnecessary to stress how painstaking and laborious such a procedure is. It should be noted that the digitizing of only one topographic map takes several hundred working hours and one or two reels of magnetic tape.

It is only natural that in this connection arises the problem of developing such a form of cartographic data representation when not only man, but also computer, should be able to read and analyse the map with comparative ease. The effectiveness of this method of solving the problem is evident: with a high reliability and speed of computer identification of symbols on such a map we shall obtain an adequate storage device. This cartographic representation of information will have the properties of a universal language, common to both man and computer, and in this way we shall improve man–computer co-operation.

For more capacious and compact data storage and for the automation of their retrieval it seems expedient to use the microfilming and holographing of maps. Information in this case can be processed not only by discrete digital computers but also by coherent optical devices, making it possible to identify and transform cartographic information, proceeding from the principles of optical filtering.

Computer-oriented maps should serve as a means to solve the problems of processing large volumes of multifarious data on natural resources. They should also assure representation of all the diversity of natural and socioeconomic phenomena, provide ample freedom in reproducing their complexity and correlativity, beginning with specific local formations and ending with the development of global processes.

The principal problem in automatic processing of maps is the identification of symbols. It is therefore necessary to develop systems of signs reliably recognized by a computer and, at the same time, easily readable by man. In this book the cartographic language complying with these requirements will be called *formalized*, and the maps using it will be called *normalized*.

The problem of developing a formalized language can only be solved by introducing the necessary modifications into the whole technological complex of map-making. Integrated elaboration of technological processes of map design, compilation, and printing, together with the development of hardware facilities and algorithms for automatic reading of images, is the most appropriate approach to solving this complicated problem.

Graphical information is usually identified in a two-dimensional space by its pictorial image (most often by the contour of a figure). The logical mathematical identification procedure is in most cases performed by a processor, and the function of the reading device is to convert graphical images into the numerical code of a computer. Not infrequently most of the computer time is spent on identifying images, and not on solving the research problems for which this identification is performed.

A study of this problem has shown that in order to ensure a higher efficiency of the cartographic data automatic processing it is necessary that the methods used to represent information, on the one hand, and the reading devices, on the other, should make it possible to identify data directly in the course of one-line scanning, i.e. in a unidimensional space.

The author's investigations have shown that, if the form of data representation on a map permits reading and identification in a unidimensional space, this will make possible:

1. simultaneous reading and analysis of several maps with the data processed on a computer or a specialized calculator (e.g. a correlator) in the on-line mode;
2. creation of relatively simple identification algorithms for a processor or a reading device freeing the processor completely from its identification functions;
3. performance of various cartometric jobs, e.g. measuring the lengths of lines, areas, etc., on a minicomputer.

That is why in developing new methods of data representation and automatic readout the possibility of cartographic data identification in a unidimensional space should be provided for. The principal functions of identification should be reduced to selective readout of data by their qualitative characteristics and some individual quantitative parameters of the model in the course of scanning.

For normalized maps as large-capacity information accumulators it is necessary to calculate optimum requirements with respect to the development of a detailed and accurate form of data representation. To this end it is necessary:

1. to establish graphical and technological requirements on the detailedness of image construction;
2. to tie up mathematically, graphically, and psychologically the detailedness and the degree of generalization with the scale of maps;
3. to formulate, irrespective of the content of data, unified requirements on the form of data representation, both graphically and technically, making it easy to read and analyse with the help of a single standard complex of specialized automatic devices and software.

The method of raster discretization (MRD) is proposed as the basic method of representing geographical data, meeting the requirements of a formalized cartographic language.

MRD is characterized not only by a simple procedure of the automatic identification of symbols (even in the presence of a topographic base) but also by the simplicity of programmes for the plotting of raster maps, as well as sufficient visual clarity with which the mapped phenomena are represented [110, 144].

The method used to create normalized maps fully preserving some traditional images is that of latent optical coding (MLC) employing the luminescence properties. Development of normalized isoline maps and stereoscopic maps with a latent optical code appears to be the most promising [111, 120].

These methods of data representation make it possible (a) to design normalized maps very similar in appearance to the traditional ones, with the existing unification of the legend fully preserved if necessary (e.g. on geological maps); and (b) to develop new types of maps with a higher degree of detail and accuracy, suited for new methods of automatic geographical data analysis.

It should be noted that a normalized map can carry not only graphical data but at the same time digital data, that can be automatically read into a computer, with the digital recording on a map presented in the form of normalized digits easily readable by man.

Formalization of the map language is a broad concept including a vast variety of representation techniques and the corresponding systems of symbols. It should be understood that formalization of language (as well as the narrower concept, following from it, of the normalization of the map form) is not limited to the above-mentioned methods only. Much remains to be done in this field, mainly in specific branches of science and industry. Normalization, which also implies standardization of the system of signs, has to be performed in accordance with the subject-matter of specialized maps. It is only necessary to formulate some basic requirements to the normalization of the most typical images for thematic and special maps.

1. On maps showing the contours of objects (qualitative background, area patterns, etc.) their boundary lines should be black. Only a few exceptions are permissible, for example when the contours are given without filling-in (inside colouring), i.e. qualitative differences are shown by the colour of the contour lines. The boundary line must be continuous or dashed at small intervals.
2. The topographic base of a thematic or a special map and the inscriptions should preferably be given in a grey, or similar, colour. The base for these maps performs auxiliary functions, i.e. it is not basically intended for automatic reading and processing.
3. Special (thematic) elements of the map content are shown by chromatic colours, preferably in a limited number—not more than five or six, dispersed as far as possible within the spectrum.
4. Linear objects (tectonic faults, rivers, roads, etc.) are shown by a continuous or a dashed line with small intervals, with the qualitative differences between them indicated by colour, thickness, or with the help of different patterns in the form of isolated bergstrichs and their combinations.

Formation of our concept is not accidental, it is closely associated with modern trends in the development of technical sciences and, primarily, engineering cybernetics. It is to the latter that we owe the word 'normalization', with its scientific content. In present-day instrument and mechanical engineering, normalization of optical, electronic, and other instruments and systems, and in documentation science and computer engineering—normalization of documents (including those presented in graphical form) is one of the most pressing problems. Specialists concerned with the methods of data input into a computer are well aware of the difficulties arising in this field, the most complex and weakest link in the general system of automatic data processing, especially with respect to graphical data. Normalization in cartography should be understood not only as the principles (procedures) of designing maps of a new type, meant to be read and processed by both man and computer, but also as the techniques for normalizing traditional maps, documents of field surveys, schematic maps, deciphered aerospace photographs, and other materials, prepared manually for computer readout in creating digital data banks and for carrying out various kinds of research [124, 131]. The principles of normalization for newly designed maps and the traditional ones already in existence have much in common and, at the same time, many features that are in principle different. One is free to choose the methods of representation and the system of symbols for newly designed maps, which considerably simplifies the solution of the problem; whereas when normalizing the traditional maps one has to adjust them to suit the possibilities of a computer in order to simplify the symbol identification function by supplying these maps with various graphical marks (codes).

Is it so necessary to attach much importance to the visual readability and clarity of a normalized map, as it is for a traditional one, if we take into account the fact that the readout and processing of cartographical information with simultaneous identification of objects will be performed by automatic devices whose detailedness, accuracy, and speed are very much higher than those of visual perception? To a certain degree the visual clarity of a normalized map should indeed be preserved. Cartographic modelling constitutes a higher specific field of modelling in general, where an indispensable requirement is to maintain direct communication between man and model, dictated by the need in spatial orientation and overall perception of studied phenomena often having a complex occurrence. This communication is necessary to assign on the map the region where a certain class of objects has to be read out. This communication is unavoidable since one cannot dispense with the operations of man coming into contact with computer when controlling the cybernetic system. The basic aim of these operations is to control a cybernetic model by means of which one will ultimately arrive at the results presented either in a digital form, e.g. the correlation coefficient, or a graphical form, e.g. a forecast map.

To deprive a map of visual clarity, that is, in essence, to detach the process of model visual analysis, is to deprive man of complete, integral and graphic comprehension of the essence of process development. One can say that it is the visual clarity where the principal merit of a map as a carrier of geographical information lies.

As a carrier of geographical information a normalized map differs favourably from conventional computer information carriers (e.g. magnetic tape) in the following respects:

1. habitual and clear data perception,
2. accessibility to all kinds of consumers,
3. possibility of simplifying the interaction of man and computer when processing geographical data,
4. possibility of selecting at will the information assigned,
5. possibility of prompt and unlimited duplication on series-produced printing equipment,
6. low cost of the carrier,
7. its relatively high information capacity (especially if the maps are microfilmed).

1.2. The Essence of Cartographic Representation of Geographical Data

Although at the basis of the cartographic form of data representation lies the language of graphics, cartography widely uses the language of mathematics, incorporating both of them in itself. For this very reason, cartographic data representation can be regarded from the mathematical and the pictorial aspects. The pictorial aspect consists in performing communicative functions based on visual, and recently also computerized, perception of the information represented. The mathematical aspect is decisive, as it is the one that ensures the projection of the represented data on a mathematically defined surface, quantitative description of objects, their precise localization and spatial distribution, together with the characteristics of density, intensity, etc.

In going from an object to its cartographic representation the process may be regarded as conversion of data on the object into the cartographic form of its description (interpretation). The interval of relationships between an object and its representation on a map is quite substantial, beginning with the object representation almost coinciding in its geometrical similarity and ending with its representation in the form of a dot.

The represented geographical data may have several qualitative and quantitative characteristics. We shall divide them conventionally into 'external' and 'internal' ones.

The external characteristics (parameters) are those presented in geodetic space, and they can be described by means of geodetic metric. *The geodetic space* is a space where the parameters of objects are represented by a function of from one to three variables, according to the general or particular scale and the projection of a map.

The meaning of geodetic metric in this case does not depend on the system of coordinates used (geodetic, astronomical, or geographical).

The internal parameters are those represented by another metric and belonging to a certain free space designated as such. *The free space* is a space where parameters of objects are represented by a function of one, two and/or *n* variables, irrespective of the scale and projection of the map. For instance, populated localities are represented in the geodetic space, and their populations, conveyed by the area of a symbol, are represented in the free space; or, for example, rivers are represented in the geodetic space (their length) and their flow rate and the volume of water masses are represented by conventional thickness of lines or the brightness of their colouring in the free space.

We shall classify all the data on objects or phenomena from the point of their distribution in the geodetic space, as discrete, continuous–discrete, and continuous. *Discrete data* characterize the objects with dispersed distribution in the geodetic space, localized over an area or at certain points on a surface. Among these data are: distribution of population and populated localities, classified by the number of population (represented in the free space); industrial facilities, classified by their capacity (represented in the free space); as well as different minor natural objects, represented to the corresponding scales as dots because of their small size. In their general form these data can be presented as a step function in space.

Continuous–discrete data characterize objects or phenomena that can be distributed in the geodetic space in the form of limited or partially limited domains with one, two, three, or many interconnections. Among objects represented by limited interconnected domains are: boundaries of the occurrences of rocks, classified by their lithological composition, age, depth, and other parameters (represented in the free space); forests, classified by tree species, estimated productivity, density of forest stand, and other parameters (represented in the free space). In the simplest case these phenomena form a limited closed domain, that can be written out mathematically as follows:

$$
\left.
\begin{array}{ll}
\text{for } \mathbf{F}_1(x,\ y) < 0 & Z = a_1 \\
\mathbf{F}_2(x,\ y) < 0 & Z = a_2 \\
\cdot\ \cdot\ \cdot\ \cdot\ \cdot\ \cdot\ \cdot\ \cdot\ \cdot \\
\mathbf{F}_n(x,\ y) < 0 & Z = a_n
\end{array}
\right\}
\qquad (1.1)
$$

where $a_1, a_2, \ldots, a_n$ are qualitative and quantitative characteristics of an object, and $\mathbf{F}_i(x,\ y) = 0$ $(i = 1, 2, \ldots, n)$ is the equation of the boundaries of its determined domain. Continuous–discrete data also include extended objects, such as rivers, geological fractures, ravines, and gullies. Although these objects belong to limited interconnected areas their specific feature is that in small-scale mapping their width cannot be represented to scale. This requires special techniques of representation. Henceforth these objects will be referred to as linear.

Continuous data characterize the objects or phenomena whose domains can be represented in the geodetic space by some two-dimensional surface in the form of continuous function $z(x,\ y)$. The domain of the function has the shape of a limited area with one interconnection and no discontinuities. Among such objects are: the gravitational field, magnetic field, and various climatic phenomena.

Representation of external characteristics (parameters) of objects in the geodetic space is, essentially, geometric conversion of data with the representation function, U, having not more than three variables, $U = F (B, L, H)$. When there is a number of internal parameters, $N_1, N_2, \ldots, N_n$, given in the free space, representation becomes multiparametric, constituting a function of $(n + 3)$ variables,

$$U = F (B, L, H, N_1, N_2, \ldots, N_n) \tag{1.2}$$

The nature of data distribution in the geodetic space—discrete, continuous–discrete, or continuous, number of internal parameters of objects and the character of their free space and metric—determine the choice and modification of the representation techniques.

All the known representation methods have a certain degree of discreteness. It is extremely difficult to achieve a fully continuous data representation by cartographic means with the necessary accuracy.

The isoline technique belongs to continuous–discrete data representation. The sign technique with a stepwise scale—to discrete–dotted representation, a diagram with a scale of intervals—discrete–discrete representation, and continuous schematic map—discrete–continuous representation.

The discrete form of data representation is, to a certain extent, enforced on us, but this should not be regarded as a shortcoming. As modern achievements in computer engineering and cybernetics have shown, the discrete form is the best not only for data representation but also for their processing. That is why the methods of developing normalized maps also use discretization techniques.

When one chooses a certain method the most important thing is the qualitative and quantitative authenticity and accuracy of data representation, on the one hand, and intelligibility for man, and for normalized maps—also for computer, on the other. In the first case the mathematical aspect is the basic one, in the second it is the graphical aspect. It is not difficult to see that the mathematical aspect is decisive. Let us consider as an example the isoline method. Irrespective of the graphical aspects (application of different scales of layerwise colouring for elevation steps), the mathematical essence of the method remains the same. Alteration of the graphical aspect, e.g. the colour of the scale or the isolines, does not change the character of the distribution of phenomena being represented. The same is true for the methods of areas (areal patterns), schematic maps, etc. No matter what graphical system is used in these methods, their mathematical substance will remain unchanged and, therefore, the nature of the distribution of objects or phenomena will stay invariable.

The graphical aspect of a method is only a means of graphically coding the qualitative differences and parameters of objects, in accordance with the preassigned representation method, for transmission via communication channels: for man—the channels of visual perception, for computer—optical and electronic channels. At the same time the graphical aspect plays a very important communicative role. The effectiveness of cartographic data analysis in solving all kinds of problems, both visually and by computer means, greatly depends on it. The graphical aspect in representation includes the problem of developing a system of signs. This system should be simple and logical, ensuring a ready perception and lasting memorization by man, and for normalized maps, prompt and reliable identification by a computer. It is for this very reason that the problem of designing sign systems is important in itself, although it is closely connected with the mathematical aspect of cartographic representation. Representation methods constitute one of the most important parts of cartographic theory. Their mathematical content is constant, independent of the qualitative characteristics of the represented data. The choice of a representation method is primarily determined by the nature of the spatial distribution of objects and the purpose of the map. In choosing the method of representation one certainly has to take into account the peculiarities of map content, the required accuracy of data representation, and how the information will be used. When choosing the representation methods and describing their techniques we shall therefore proceed from the nature of the spatial distribution of data and the purpose of the map. By their purpose we shall subdivide maps into two categories:

1. popular science and educational maps;
2. scientific reference and specialized (topographic, specialized, thematic) maps.

For popular science and educational maps most important is graphically clear data representation meant exclusively for visual perception. The graphical aspect in them comes into the foreground and the mathematical aspect, as it were, stays behind.

In the problem of designing normalized maps, of primary importance is the development of new methods for the representation of data, including the mathematical, graphical, technical, and technological aspects, for automatic processing. No doubt some traditional representation methods (e.g. the isoline method, the method of qualitative background) will essentially remain suitable for normalized maps as well. However, to raise the reliability of automatic identification of symbols and increase the information content and accuracy of represented data, it is necessary to carry out research and development with respect to both the graphical design of maps and the mathematical and technical aspects of representation, including the elaboration of innovative methods. These ideas, first formulated in references 94 and 95, have been convincingly confirmed in a number of experiments. Proceeding from the point of view that the higher the communication content of the map form the easier it yields itself to formalization and automatic processing (including its reading), we shall formulate the general requirements on the new methods of data representation (MDR):

1. MDR should make possible:
   (a) reliable automatic identification of symbols;

(b) effective (technically and economically) automatic selective readout of cartographic data.

2. The graphical form of MDR implementation should include an easily memorizable and logical system of symbols assuring clear and vivid data perception by the human visual analyser.

3. MDR should permit transmitting the data:
    (a) of enhanced information capacity as compared to that of traditional cartography;
    (b) of enhanced accuracy and authenticity.

1.3. The Essence of Cartographic Generalization

The principal requirement in cartographic representation is the spatial conformity of the qualitative and quantitative parameters of objects and phenomena to their actual distribution. Of major importance on large-scale maps is the requirement of accurate representation of metric parameters of objects and their outward geometrical likeness (accurate representation of boundaries, areas, extent of objects, etc.). On small-scale maps these requirements in most cases prove to be impracticable because of generalizations entailing the diminution of images. Consequently, a different approach is needed here for the representation of these parameters.

The long-standing generalization techniques for traditional maps have a tendency of maintaining the 'natural' similarity in the representation of objects. For instance, contours of forests that cannot be shown to scale on a given map are exaggerated, merged, but are shown within their 'natural' boundaries, or else discarded. This results in a serious violation of the conformity of represented data to reality, affecting all the parameters: contour boundaries do not correspond to their actual location, the number of contours does not coincide with their actual number, contour areas differ from their actual size, the weighted centres of contours are displaced from their actual location. As a result, a map carries erroneous and therefore harmful information, misleading and misinforming the user. Such a map does not yield the necessary reliable data on either of the above-mentioned parameters which could be used in analysis to solve any serious problems. The traditional technique can only be acceptable for popular and educational maps where the outward pictorial image, clearly approximating the distribution of objects or phenomena, will be sufficient, as well as for general maps used for special purposes.

Such generalization technique is usually justified on the grounds that by knowing the natural peculiarities of the distribution of certain phenomena and their interrelation one can correctly generalize and mark boundaries of the represented object or phenomenon. It is possible to agree with this, but only for large-scale maps, and with a serious reservation. Not by chance there exists an opinion that it is difficult to find two cartographers who would perform the generalization of the outlines of any substantial portion of a map in a completely identical way. And this is quite understandable because such a generalization technique, regardless of how much was written about it, will remain basically subjective, as it proceeds from the individual approach of a cartographer, his personal understanding of the interrelations between the mapped phenomena (which as yet may have been poorly investigated). It goes without saying that in this case one cannot expect any identity of generalization results. And the user of the map is, naturally, unable to penetrate into the thoughts of its designer, his intentions in generalizing a certain contour or selecting the objects, so as to introduce the necessary corrections in the results of readout. Subjectivity of such generalization is aggravated by the lack of common views and concepts regarding the interrelation of phenomena (especially when these are very complex). It is also obvious that the principle of representation, consisting in the preservation of the outward geometrical similarity by searching for some 'correct' and 'natural' boundaries, is unable to produce conformity to reality (especially on small-scale maps) even in the basic parameters.

Taking into account the aforesaid, we shall have to say that maps designed for scientific and industrial purposes call for a different approach to generalization [24, 52, 117, 136, 148, 151, 152]. It is the formalized approach to the problem that can provide objectivity and authenticity of data representation. For representation to a scale that does not permit showing the natural outlines of objects, formal transformations of conventional similarity (quasi-similarity) are needed. These, although giving only an approximate outward likeness, are able to represent the main parameters authentically and factographically, i.e. to perform a separate representation of the parameters of objects on several maps or in admissible combinations on one map. For instance, a river network can be represented with sufficient authenticity on medium- and small-scale maps by its following parameters:

1. meandering (by inflection points, density of meanders in isolines, etc.);
2. length (by points along the channel at equal intervals);
3. width and water discharge (multiscale representation along the width of the river with the image brightness gradation);
4. volume and area (by points along the channel at equal intervals and multiscale representation along the width of the river with the image brightness gradation);
5. length of small rivers (by multiscale dashes along the course).

Intersection of symbols caused by the overlapping of objects which results from generalization can be actualized by superimposing raster unidirectional lines in accordance with MRD. This very course of generalization lies at the basis of the representation methods discussed in Chapter 2. Since the methods of data representation include data transformation processes, they are closely interrelated with the principles of generalization. Moreover, these methods are devised to assure the possibility of using computers and automatic devices for the reading and plotting of cartographic data.

Generalization should provide for the restoration of the exact parameters of objects (possibly, by making corrections). This approach is more valuable than even a good (from the geographical point of view) generalization that does not give us an opportunity to take into account the variation of the quantitative parameters of objects. In this case mathematically substantiated generalization principles are needed. It is possible to develop a technique of generalization at several levels, with the use of several formalized filters. For instance, at the first level a purely formal filter can be used to select and generalize objects, proceeding from preassigned criteria, applying different approximating functions, etc. The second level should be of a higher order and take account of the interrelation of objects and the peculiarities of their spatial distribution and development. Owing to the formal unambiguity and definiteness of such generalization filters one will be able, when using the map, to take into account the level of filtering and in this way introduce corrections into results of map reading.

Cartographic generalization, depending on how the represented geographical data are to be used, may be subdivided into two principal types: scaled generalization and special-purpose generalization.

Scaled generalization serves the main purpose of compressing the scale of data, with their content, form, and purpose remaining unchanged. Most typically, scaled generalization manifests itself in designing topographic and analytical maps. Since losses of data are inevitable in going to smaller scales, the problem arises of isolating and representing what is the most typical, basic, and essential, because mechanical omission of small details may result in substantial information losses. One should acknowledge that if in going to maps of a smaller scale, even when optimal generalization techniques are applied, there are losses of data, it is a negative fact. Unfortunately this is inevitable. It is necessary to search for and find such methods and techniques of data compression that would result in minimum information losses, so that the data on the map should be sufficiently accurate, authentic, and have an optimal detailedness. This should certainly be done with due account for the requirements imposed on a map as a graphic carrier of information. In the search for such methods and techniques one should not be restricted only by selecting and modifying the outward geometry of cartographic image.

Special-purpose generalization has narrow aims. This is caused by the specificity of the use of represented data. This generalization is not a function of the scale, and is mainly applied in studying objects or phenomena by designing derivative maps, depicting the most general, typical, essential aspects of data—i.e. those aspects or parameters that serve the purpose of their use. It is thus a purposeful transformation of data according to the subject-matter of research and consisting in isolating the typical and essential, and ascertaining the most general interrelations, notions, and categories. The most typical example of special-purpose generalization is the transformation of data concerning the relief (topographical maps), the geophysical fields, ore bodies, geochemical elements, etc., into trend surfaces characterizing the most typical general distribution of these data, governed by a certain degree of regularity, and illustrating the basic tendencies in the localization of the objects and phenomena in question. Another typical example of special-purpose generalization are the transformations associated with designing maps of the distribution density (occurrence) of objects or phenomena, proceeding from certain assigned parameters, e.g. designing maps of population density, woodiness, and river network density (length, meandering).

1.3.1. Scaled generalization

Scaled generalization is not a simple mechanical data compression but a complex process, including logical transformations. There are three forms of scaled generalization: formal, logical (logical mathematical), and multilevel (quasi-similar).

Formal generalization consists in the general filtering and smoothing out of data according to criteria calculated proceeding from strictly formal rules which provide for visual perception of the image and the possibility of using technical devices for its reproduction and processing. This form of generalization includes the interpolation and equidistant methods in automatic generalization of the boundaries of contours and linear objects. Selection is also performed based on formally assigned criteria. These methods take little account of the peculiarities and significance of the represented geographical data.

Logical (logical mathematical) generalization consists in the logical mathematical data transformation with the help of differentiated criteria formulated with due account for the peculiarities, substance, and significance of represented objects or phenomena. Its procedures have a logical mathematical meaning and are aimed at a clear representation of the typical and significant, with the characteristic features of the mapped object being singled out. Of the methods of automatic generalization the closest to this form is the structural method, where the procedure of smoothing out linear elements is performed based on census criteria and taking into account the parameters of the portions of the curve assigned by its characteristic points. All this is done in a differentiated way, depending on the peculiarities and significance of the represented objects.

Multilevel (quasi-similar) generalization consists in a quasi-similar transformation of data aimed at compressing them by means of differentiated representation of unplottable elements in other larger scales. Its main objective is to achieve minimum information losses by means of different graphomathematical methods of its compression. This objective is reached by quasi-similar transformations with a multiform data representation for visual perception. This implies the possibility of singling out what is typical, significant, and general by graphical means, using the brightness, colour, and texture characteristics of an image, as well as the raster structures. The idea of multilevel generalization lies in splitting the data into several levels of detailedness and significance with a different degree of generalization and using the representation techniques assuring a graphical and topological possibility of representing both small details, lying beyond the limits of possibilities provided by the scale of the map, and large ones, which are for the most part the typical, principal, and essential. An important aspect of multilevel generalization is the characteristic feature of the graphic representation of information, when it is visually perceived with a different degree of prominence: in the foreground are identified the most generalized data, then come the less generalized and, in the background the least generalized ones. This trend in generalization appears to be the most optimal in many respects, primarily in that multilevel generalization is associated with minimum information losses, with the clarity of images, providing a clear distinction between the primary and the secondary, the general and the particular. A concrete example of multilevel generalization is given in the section on the methods of multiscale representation (see Chapter 2) [110, 117].

However, in all methods of scaled generalization it is impossible to avoid information losses in going to smaller scales. It is also clear that the more radical the selection the greater the losses. A quantity of information dP on an element of area $dS = dx \times dy$ will be referred to as information content of a map in a certain point.

We shall define scaled generalization as a decrease in the information content of a map, dP/dS, and regard it as a function of scale variation m and, in some cases, of the type of projection, at a constant content and system of symbols.

The modulus of difference

$$\left| \frac{dP_2}{dS} - \frac{dP_1}{dS} \right| = g \tag{1.3}$$

where subscripts 1, 2 refer to the designed (generalized) and the original map, respectively, will be called the scaled generalization effect.

Estimation of scaled generalization as a process, reducing the information content of a map, should be considered only relative to the actual surface element, then the generalization effect will increase as one goes to smaller scales of the compiled map and, consequently, information losses will increase.

Especially good prospects open up for scaled generalization in designing maps with latent information. The data on a normalized map can be divided into two parts: one to be read by man, and the other only by computer; or one for man and computer, and the other only for computer. The first part of information must necessarily assure visual perception by man, while the other must correspond only to the resolution power and the logical capability of a computer.

In future, based on special materials and using computers and special devices for data plotting, we shall be able to produce maps with a high information capacity. Potentialities of computers are increasing with every passing year. Even today a computer in many ways surpasses the visual analyser of man. There is no doubt, therefore, that the development of geographical data media in the form of maps containing easily readable, clear, graphically generalized information for man; and detailed, accurate, highly informative, and readily decipherable data for a computer, has a great future.

1.3.2. Special-purpose generalization

Special-purpose generalization involves construction of integrally generalized surfaces, representing a certain form of abstraction, i.e. a specific type of generalization identifying the most general regularities of a phenomenon, ascertaining the region of the spatial distribution of discrete objects, shown in the form of isolines, areas, etc., with the purpose of studying interrelations, distribution characteristics, etc. Special-purpose generalization is primarily necessary where research is performed [108, 120].

The generalization process is essentially that of data transformation, most clearly manifested in special-purpose generalization.

Data transformation, with the aim of ascertaining what is the most general and significant, is successfully achieved by approximating surfaces with the help of algebraic and trigonometric functions; for example in decomposing surfaces presented on a map into isolines. Such decomposition of surfaces is performed to single out the principal factor forming the structure, the most essential typical features of a given phenomenon, such as ascertainment of zonal climatic regularities by excluding the local peculiarities. Maps representing such surfaces are called trend maps in geology. When analysing the structure of crystalline bodies and identifying ore-bearing sites the plotting of geochemical trend maps proves to be a very effective method of analysis. These maps, characterizing the distribution of the quantity of a certain element in rocks, make it possible to reveal the basic trends in the location of different ores. Approximation methods have been known for a long time and are used to transform gravitational and magnetic fields. With their help one can exclude small elements of local and casual nature and clearly isolate anomalies caused by larger tectonic units. It is true that sometimes secondary details against the background of the main structures are of particular interest; for example, identification of local variations of gravitational field anomalies against the background of regional changes. These methods of surface decomposition initially developed in geophysics are now widely used in geography and geology, especially in structural–geomorphological analysis and tectonic studies, e.g. in studying local neotectonic movements against the background of general oscillatory ones, in analysing the regional behaviour of planation surfaces and their local deformations.

Two types of trend analysis can be identified. The first one using weighted moving averages, the second approximating the observations with an analytically given function of independent variables. The most widely used apparatus of approximation is based on choosing a polynomial of the corresponding power with the help of the least-squares method.

Depending on the concrete objectives of research and the approximation method applied, different procedures can be used to find the trend surface. The trend surface must be approximated by a function of coordinates $Z = f(x, y)$. We shall now briefly describe some of the presently known procedures.

One of these generally used is the approximation with the help of nonorthogonal algebraic polynomials. A polynomial of two variables is most often taken as the approximating function:

$$Z = a_{00} + a_{10}x + a_{01}y + a_{11}xy + a_{20}x^2 + a_{02}y^2 + a_{30}x^3 + a_{21}x^2y + $$
$$+ a_{12}xy^2 + a_{03}y^3 + \cdots \tag{1.4}$$

The problem of finding the approximating surface is reduced to determining coefficients $a_{00}, a_{10}, a_{02}, a_{11}, a_{20}, \ldots.$ To calculate the values of coefficients, a sequence of Z_i values is taken from the original map, and then the least-squares method is applied. The number of coefficients to be determined depends upon the power of the polynomial. This dependence is expressed by the following formula:

$$N = \frac{(n + 1)(n + 2)}{2} \tag{1.5}$$

where N is the number of coefficients, n is the power of the polynomial.

Special-purpose generalization is inapplicable to analytical and topographic maps having a wide variety of users, when it is impossible to determine in advance for what purposes the maps will be employed. It can only be applied for a map with a strictly determined purpose.

If a map is used by an engineering designer to plot a railway line, what is important for him is the accuracy and reliability; he does not need data distorted by the generalization of the individual forms of relief, e.g. on a topographic map, or of the geological contours on an engineering–geology map. At the same time for a geomorphologist or a geologist, when studying tectonics, it is most important to identify interrelations and regularities; that is why generalization of the individual forms of relief, emphasizing its structure, is of great positive significance.

Thus in designing popular, educational, and other maps for a wide variety of users, a most important factor is the graphic, easily comprehensible representation of data resulting from an increased generalization effect and aesthetic characteristics of a map. In designing maps for scientific and industrial usage, when accurate and detailed representation of geographical data is required, scaled generalization, with restrictions imposed by the technological printing equipment and psychophysiology of image perception, is applied. In designing maps meant for solving research problems, special-purpose generalization is used.

1.4. The Essence of Cartographic Information

Recent developments in such modern sciences as automatics, computer engineering and information science have had a strong influence on traditional engineering sciences, including cartography. The use of automation in cartography has required a radical revision of long-standing concepts and ideas. The notions customarily applied in traditional cartography were 'cartographic image', 'legend', 'conventional signs', etc. Today, along with these, such concepts as 'cartographic data', 'system of symbols', 'readout and coding of data', have come into use. The appearance of new terms and concepts is associated with the necessity of operating with them in conducting various automatic processes, primarily in designing and analysing maps. The traditionally established concept is that everything associated with cartographic information, mapping, or cartographic image is embodied in a map serving as a model, a graphical document representing, usually on paper, certain information concerning the Earth, directly perceived by eye. When automatic devices and computers began to be used it turned out that cartographic information could be transformed, recorded, and stored on other information carriers, such as magnetic drum, magnetic tape, punched tape, hologram, that do not have the customary appearance of a map—a drawing of the Earth's surface enclosed within a certain frame. Nevertheless, after the transformation the map does not disappear, it is incorporated into these carriers with all its imagery, including colour signs, legends, etc., with the only difference that the image is represented in the form of coded electrical, optical, and other signals. Whenever necessary the map can again be converted into a visual image on the screen of a cathode-ray tube (when this conversion is performed electronically) or in the form of a hologram (when the conversion is performed optically). In this case one can display at will not only the whole map, but any of its portions with a different content, and even its individual elements, that is the image of individual objects such as a river, road, or populated locality. These possibilities certainly depend on how the images on the map have been coded. If the image of only one element, for example a river (with permissible distortions), is taken out of computer memory and displayed on a screen, can it be regarded as a cartographic image? Undoubtedly yes, because all the points of this image are represented on a mathematically determined surface, that is in a certain cartographic projection, and preserve the basic characteristics of the cartographic image.

From this it follows that the whole map as a graphical document should not necessarily be regarded as geographical information *per se*. It can be a part of data, read out from the map, converted into electrical signals, and recorded in a computer data-storage medium, to describe the image of a certain element or object on the map (a lake, river, road, populated locality), or converted into optical signals and represented by a hologram. These individual elements (images) should not necessarily belong to a certain map or several maps. They may lie beyond the traditional concept of a map. For instance, the delineation of large tectonic structures or aircraft and spacecraft flight routes can be globally applied to the whole planet or its major regions. They may be represented in a form which is unusual for a map, e.g. in the form of narrow bands, or simply as schemes of an arbitrary pattern.

With such an approach to defining the concept of information any other carriers of cartographic data, including spatial—such as a relief map, hologram, or globe—can be regarded as information, since all the points of these images are represented on a mathematically determined surface. It can be said that a traditional map is one of the widely used data-storage media.

A traditional map on paper is also, in a sense, a storage device, but for man only. If it becomes readable by man and computer, that is if its form provides for the resolving power and logical facilities of a computer, it will then simultaneously perform the function of a computer storage, such as magnetic tape, magnetic disc, punched tape; but, unlike these, it can be directly perceived by man.

Cartographic information includes data on the properties of objects in a graphical form, but the properties of the objects themselves as information content belong to the branch of science using the cartographic form of representation. The term 'graphical form' is understood here as the form of data, their graphical code, represented by colour, geometry of an image, graphical mathematical plots, symbols, and their logical interrelations. As soon as it loses its graphical and mathematical properties inherent in the cartographic form of representation it ceases to be cartographic information, acquiring a different form, such as that of a digital model of a region, but in its content it still remains information for the branch of science to which it belongs.

Cartographic information must reveal the content of geographical data with respect to the distribution of objects or phenomena in geodetic and free space—by means of cartographic methods of representation, and with respect to the logical interrelations between their qualitative characteristics—by means of a system of cartographic symbols, irrespective of whether the purpose of the map is known. Hence the lack of knowledge on the substance (content) of data on a map is admissible, since it on no account means that the possibility of obtaining the necessary idea of the spatial distribution and the interrelations of objects or phenomena is ruled out. The situation is similar to a mathematical model describing a certain process, structure of objects, etc., which makes it in principle possible to describe and interpret these data when the concrete branch of science or technology to which they belong is unknown. Moreover, the same cartographic representation methods and systems of symbols, as well as the same mathematical techniques and symbols, can be used with equal success to describe and model information which is highly diverse in content. However, lack of knowledge of the purpose of the map certainly reduces the researcher's possibilities.

Concrete geographical data, read out from a map and recorded in digital or verbal form, containing no information on the form of the map (symbols, mathematical base, etc.), can no longer be regarded as cartographic information since it does not contain any features indicative of its belonging to a map. The same information can be obtained from the results of field surveys, statistical data, etc.

Determination of the concept of cartographic information is necessary both for the development of cartographic theory and the solution of practical problems, primarily in designing automatic cartographic systems, in various transformations of data, as well as in their storage and transmission.

It has been noted above that a map is a graphical mathematical model. By means of graphical mathematical methods it describes the objects constituting the toposphere, lithosphere, etc. If all images of the map are converted into a digital form containing all the necessary information on the form of symbols, the mathematical base, as well as all the accompanying external information of the map (such as nomenclature and legend), this kind of converted information can be called a *digital map*. The necessity of constructing digital maps is mainly associated with cartographic data transmission and some procedures of their automatic processing, especially in the automatic plotting of derivative maps. To store data in the form of digital maps in automated systems is inexpedient, with the exception of digital maps having a unified system of symbols (topographic, oceanographic, etc.). In these systems it is more justified, economically and technically, to store data in the form of *digital models* describing objects or phenomena not cartographically but by means of purely mathematical methods with digital recording of the original data that has been obtained, for example, in field surveys. It is obvious that digital models ensure more accurate and detailed information storage, since they lack the defects of a map requiring a generalized data representation because of the restrictions imposed by graphical presentation. A digital model may not contain any cartographic information. All its substantive data can belong to a specific branch of science (such as geology, geography, meteorology, or economics).

Based on digital models it is possible to design any kind of map with different scales and certainly to solve most diverse engineering-survey problems, or perform various investigations of natural and socioeconomic objects or phenomena.

Normalized cartographic information is conveniently stored (from the point of view of media capacity and efficiency of its retrieval and usage) on microfilms or in natural form. These are more readily accessible to a great variety of users then computer storage media and can be easily and cheaply duplicated by means of serially manufactured printing equipment.

From the above it is clear that cartographic information is a special form of graphical data, communicating knowledge of the Earth and celestial bodies on a mathematically determined surface (i.e. in a certain cartographic projection) and describing it graphomathematically with the help of a system of symbols in the form of visible images or coded electrical, optical, and other signals on any kind of carriers (such as magnetic tape, photographic film, or paper).

1.5. Estimation of the Quantity of Cartographic Information

Estimation of the quantity of cartographic information presents an important problem, especially in connection with the automation of cartographic procedures. When assessing the information one should proceed from the quantity of variety of map image elements (as members of a set). By variety we understand the differences between the elements of a set. If two signs of qualitative background on the map field differ only by their colour intensity, their set consists of two elements with variety. This elementary distinction, the difference between two symbols or elements of the image, is the primary unit of the quantity of information. It is self-evident that the larger the number of dissimilar elements in the image the more information it carries. In the automatic processing and transmission of cartographic data the unit of measurement is a 'bit'. If a map is read out by a scanner and the data in the form of a discrete sequence of video signals (dots quantized by brightness) are read out against a background unchanging in its colour characteristics (i.e. the background in the filled-in, solid colour, form), these video signals as elements a_i of set A, being the same, are devoid of variety. The maximum variety will be found in the case when all elements of set A differ and their probabilities $p(a_i)$ are equal. In this case the quantity of information is estimated based on the notion of probability, used to describe situations with uncertainty. The degree of uncertainty of a communication is measured by the value—usually referred to in the general theory of communication as informational entropy (H). Entropy also denotes the measure of the quantity of information. It is a function of probability distribution $p(a_i)$. In its general form it can be written as

$$H(A) = H[p(a_1), p(a_2), \ldots, p(a_n)] \qquad (1.6)$$

When the probability of the next signal does not depend on that of the previous one, entropy can be calculated from the formula:

$$H(A) = -\sum_{i=1}^{n} p(a_i) \log p(a_i) \qquad (1.7)$$

The unit of entropy depends on the chosen logarithmic base. If the base is equal to two, information is measured in bits.

Depending on the problem being solved different principles of estimating the information can be used. A stochastic approach is acceptable for problems related to information transmission via a phototelegraphic communication channel and its input into a computer with, e.g., photoelectronic devices. In these processes cartographic data are transmitted without any formalized information on the qualitative and quantitative characteristics or on the form of the symbols in their cartographic meaning, but simply as optical signals obtained discretely along the lines in the course of scanning. These signals are of different brightness in every electronic scanning channel, corresponding to a specific portion of the map. With this kind of computer input it is very important to know the quantity of input information so as to estimate the resources of computer memory.

Let us first consider the case of a one-channel scanner which gives a one-colour output with image quantization only by brightness. Every optical signal obtained by scanning the map will be regarded as a set of communications represented by the values of brightness for the elements of image discretization, n. If we take the number of brightness quantization levels equal to m (in modern scanning devices m reaches 256), the number of lines equal to k, a constant number of discretization elements in a line equal to n, then the quantity of information H, represented in the computer storage by binary digits, can reach

$$H = n \times k \times \log_2 m \tag{1.8}$$

For a multicoloured map the quantity of information is multiplied by the number of separate colours used in the map.

For the transmission of cartographic data represented in a coded form on computer storage media the stochastic approach is also applied. The data are represented at the input of a transmitting device in the form of combinations of elementary signals for every symbol (sign). Here it is very important to attain the economic efficiency of the coding system, which is determined by the average number of elementary signals $\bar{n}$ representing one symbol of a given communication in accordance with the probability of its appearance. The value of $\bar{n}$ cannot be less than entropy H:

$$\bar{n} \geqslant H = -\sum p_i \times \log_2 p_i \tag{1.9}$$

but can approximate it to any degree of closeness.

When it is necessary to determine the total quantity of cartographic information on a map with due account for its semantic aspects in the course of its compilation and analysis a different approach is necessary. One also has to bear in mind that a map is a static graphomathematical model, and therefore cartographic information is quite definite and unambiguous in space and time. In this case other methods of information assessment are more expedient.

Such methods, based on a combinatorial approach, have been described in a number of publications. It is suggested that the quantity of information should be determined by calculating combinations of qualitative, quantitative, and temporal characteristics of symbols from formula

$$I_s = \log_2 \left(\sum_{k=1}^{n} R_k N_k \prod_{i=1}^{m(k)} D_{k,i} \right) \tag{1.10}$$

where: n is the number of symbols designating the types of territorial objects, $m(k)$ is the number of the characteristics of symbols, N_k is the number of symbols denoting the object of every type, R_k is the date for each symbol of an object, $D_{k,i}$ are gradations of a characteristic of the kth form [43]. The underlying principle here is that of absolute and relative differentiation of the represented object and its semantic content.

For cartographic data, variety primarily consists in the geometry of the drawing (different curvature, length, width, and all kinds of their combinations, making up individual symbols and their complexes), in diversity of colour characteristics (brightness, hue, colour saturation), and finally in the differences in combinations and interrelations.

Best for cartographic information is the estimation of the quantity of information through the measure of complexity of a graphical drawing. Complexity of a cartographic image is associated with its following properties:

1. the number of zero curvature points at the boundary of an object (contour, discrete symbol, etc.), or the number of angles and the variation of their size;
2. the result of dividing the squared perimeter of an object by its area (this number reaches its minimum value for a circle and is equal to 4π);
3. for a coloured image—the number of brightness, hue, and colour saturation gradations.

In estimating the quantity of cartographic information it is necessary to distinguish between the complexity with respect to the drawing of symbols and the complexity of an image with respect to the spatial location of substantive data (boundaries of contours, giving a two-dimensional description of the areas of objects or phenomena distribution; isolines, giving a three-dimensional description of the spatial distribution of phenomena, etc. This is associated with the fact that an increase in the complexity of the drawing of a symbol, with the invariability of the information it conveys, decreases the information content of the symbol itself, and therefore that of the map, whereas an increase in the complexity of contour and isoline boundaries increases the information content of the map.

Thus, systems of symbols have their own information content assessment. It considerably differs from all the data related to the spatial position of objects and their quantitative parameters.

In designing a system of symbols it is necessary to satisfy two basic requirements:

1. assure the best possible perception of symbols, based on the psycho-physiological characteristics of the human visual analyser (for normalized maps—also on the logical facilities and the resolving power of a computer);
2. assure the highest possible information content of symbols.

Both these requirements are interconnected. To a certain extent they produce a mutually negative effect upon each other; i.e. with a better perception of symbols the information content decreases, and vice-versa. For instance, when the size of a symbol increases, its perception improves, but information content over the field of the map decreases. The inverse case is that of saturating a symbol with graphical patterns, that is raising the number of graphical distinctions, so as to expand the qualitative and quantitative characteristics conveyed by the symbol reduces (complicates) its perceptibility. With the problem formulated in this way, formal criteria are required to enable one to find ways of estimating and constructing an optimum system of symbols. For this purpose the concept of informational efficiency of a symbol (IES) has been introduced in reference 120.

IES is understood as the ratio of the quantity of semantic information σ (i.e. the number of substantive characteristics the symbol has) to the measure (degree) of the symbol complexity λ:

$$H_e = \frac{\sigma}{\lambda} \qquad (1.11)$$

It follows from the formula that the simpler the outline of a symbol and, at the same time, the more semantic the information it conveys, the higher the IES. The value of IES cannot be greater than unity, as the number of represented substantive characteristics cannot be greater than the quantity of variety in the drawing of a symbol, i.e. not greater than the number of differing graphical features present in the symbols and between them.

IES estimation is most expedient for designing discrete abstract symbols.

When we have to face the problem of evaluating the information content, I, of a map as a model, taking into account the whole of its content, the geographical information should be estimated by the number of qualitative and quantitative characteristics, dP, falling within an element of area, dS. Quantity dP must include: (a) the measure of complexity of boundaries for contours, linear objects, etc., as the data characterizing the position of objects in the geodetic space, dP_g, and (b) the number of the characteristics of semantic information that describes the properties of objects in the free space, dP_f:

$$dP = dP_g + dP_f \qquad (1.12)$$

Necessity of the overall estimation of the information content for a concrete map as a model mainly arises in the automated compilation and analysis of maps.

In designing a map, however, it is most important to choose or develop the optimal systems of symbols and the representation methods, yielding the highest information content and perceptibility at the preassigned content and purpose of use. In this case it is not so important to perform the general estimation of the information content of a map as to assess the possibilities of representation at points with highest concentration of objects, taking into account the number of their qualitative and quantitative characteristics to be represented.

The problem of estimating the cartographic information as a result of the interaction of information carrier (map) and information recipient (reader, map user) by counting the cartographic images, as it is presented in reference [6], contains a fundamental difference from the above-mentioned approaches to determining the quantity of information. Unfortunately, this problem, though it appears to be of interest, is basically subjective and, therefore, can have no formalized correct solution. This is explained by a not quite justified premise: the user can supposedly extract from the map more information than was used in creating it [6, 134]. If a map is regarded as an information carrier, then, on the strength of the well-known laws of cybernetics, no information carrier can yield more information than was used in producing it. In other words, the quantity of output information $h'_1(x, y), h'_2(x, y), \ldots, h'_{n-1}(x, y)$, h'_n can always be less than or equal to the input information $h_1(x, y), h_2(x, y)$, $\ldots, h_{n-1}(x, y), h_n(x, y)$ (Fig. 1.1). The above opinion is based on the understanding that, since a map is simultaneously a model, the objects and

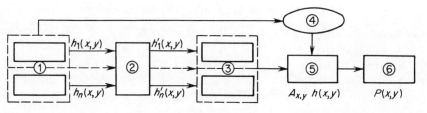

Figure 1.1

phenomena represented on it may be interrelated and governed by certain laws, of which the compiler of the map did not necessarily have to know when he created it. The user of the map, however, possessing the necessary knowledge, can reveal them. This very information, additionally revealed by the user, constitutes the information 'increment' which he supposedly extracts from the map 'over and above the information used in creating it'. That such a conclusion is unjustified is obvious. First of all, this information 'increment' is nothing else but the result of analysing and interpreting the read out information—3 from the map (maps)—2 and the information contained in the user's memory—4 as a result of life experience and specific knowledge acquired by studying the analysed objects and phenomena (Fig. 1.1). This 'increment' can also be regarded as the result $P(x, y)$ of speculative, as well as computer-assisted, data processing with the help of a certain operator $A_{x,y}h(x, y)$, represented by any of the known methods used to transform and analyse cartographic information read out from one map or a series of maps. The information 'increment' obtained in this way by no means shows that the information extracted from a cartographic model can be wider and richer than that implanted in the map by its compiler. The same information 'increment' can also be obtained from the analysis of an aerial photograph, a photograph, a landscape, a book, and many other sources. The magnitude of this information 'increment' mainly depends on the erudition of the map user. It is also quite obvious that estimation of information by counting the cartographic images is practically inapplicable, since 'from one and the same map different readers using different technical means are able to perceive a different quantity of cartographic images'. It is impossible likewise to estimate the constant objectively existing quantity of cartographic images (CI) from which one can estimate the quantity of cartographic data (CD), just as it is impossible to know everything about the interconnections, relations, regularities, etc. characterizing the objects and phenomena that can be represented on a map with the help of signs (S). At the same time, such an approach to estimating the information contained in a map, where combinations of individual signs and the relations between groups of signs are considered with the help of the three-membered formula S–CI–CD [6] is of theoretical interest in developing further the cartographic communication.

1.6. The Essence of Cartographic Modelling

Cartographic modelling comprises the processes of designing and compiling maps, which consist in constructing, with the help of graphical devices and mathematical methods, a cartographic model of objects or phenomena on the geosphere and celestial bodies, and interpreting the model with the help of graphicomathematical methods of cartographic data transformation.

The basic components of a cartographic model are:

1. *The mathematical base* providing for a certain scale of the model and the representation of the surface of the Earth ellipsoid on a plane within the assigned territorial limits and establishing a strict one-to-one correspondence between the coordinates of points on the model and in nature.
2. *The methods of geographical data representation* describing the spatial distribution of geographical data and establishing the correspondence between the parameters of represented objects and phenomena on the model and in nature.
3. *The system of symbols* ensuring the transmission of represented geographical data via the visual channel to man and optical electronic channel to a computer.
4. *The content of the map*—geographical data. The geographical data are 'supplied' by such branches of science as geology, geography, geophysics, astronomy, and economics.

In terms of the present-day theory of modelling, cartographic modelling can be classified as ideal modelling. It combines in itself the properties of graphic, symbolic, and mathematical ideal modelling. Depending upon the type of cartographic model, the procedures of its designing can, to a greater or lesser extent, approximate one of them. Thus, for example, closest to graphic modelling are the procedures of constructing tourist and display maps; closest to mathematical modelling are prognostic maps and maps of isocorrelates.

The graphic form being correct for all the three types of ideal modelling, and taking into account the indispensable presence of mathematical modelling procedures in the constructing of a cartographic model, a map can be described as a graphomathematical model [120].

A map as a model can simultaneously be both gnosiological and informational. The *gnosiological essence* of a map as a graphomathematical model lies in the possibility of using it, in conjunction with special methods of investigation, to establish the objective laws governing the development of natural and social processes. The *informative essence* of a map as a model lies in its ability to store the data concerning objects or phenomena in the form of graphical patterns.

A number of works have been devoted to defining the map as a model and to the concept of cartographic modelling. Their analysis reveals that the problem requires further study and development.

Two approaches to defining a cartographic model are suggested by the author:

1. A cartographic model S_i is understood as representing on a plane a part of the Earth ellipsoid surface in projection $x = f_1(\varphi, \lambda)$, $y = f_2(\varphi, \lambda)$ and to scale m_i and as the corresponding representation of qualitative and quantitative parameters of geographical data, $N_1, N_2, \ldots, N_n$, which are a function of spatial coordinates x, y, H and time t, with a certain degree of their filtering and generalization g_i,

$$S_i = m_i F[N_1(x, y, H, t), N_2(x, y, H, t), \ldots, N_n(x, y, H, t), g_i(x, y, H, t, m_i)]$$
(1.13)

Parameters N_i can constitute the characteristics of one or several phenomena, e.g. the characteristics of a forest: N_1—tree-stand composition, N_2—tree-stand density, N_3—maturity of tree-stand, etc. They can also include the elements of topographic base. The communicative aspect of the representation of these parameters can be expressed as the function $N_i = G[p_i(x, y)]$, where p_i represents certain graphical parameters (colour saturation, density or thickness of lines, area of a symbol, etc.) corresponding to their quantitative and qualitative characteristics.

2. A cartographic model is understood as the union of sets of substantive and communicative elements with the relationship of the order and similarity with the objects of modelling assigned on it. These are represented in a graphical form, providing an integral and clear visual perception of interrelated phenomena and objects

$$K = \{A \cup B \cup C; \alpha, \beta\}$$
(1.14)

where K is the cartographic model (map); A is the set of substantive thematic elements; B is the set of mathematical elements; C is the set of communicative elements; α is the similarity relation; and β is the order relation.

Relations of orders between the elements of sets A, B, C and of similarity with the objects of mapping are determined in the course of modelling, which can be expressed as

$$M_k = \{(G : K_i - K_d), R\}$$
(1.15)

where K_i is the initial cartographic model or, when different sources are used, the set of initial data elements; K_d is the developed model (map); G is the transformation operator; R designates the rules for handling the model (the use of the map). The elements of set A are: a_1—purpose of the map, a_2—substantive elements of the mathematical base, a_3—substantive elements of the geodetic base, a_4—topographic substantive elements, a_5—thematic substantive elements, i.e.

$$A = \bigcup_{i=1}^{5} a_i.$$

The elements of set B are: b_1—projection of the ellipsoid surface onto a plane,

b_2—methods of geographical data representation in accordance with the assigned projection, b_3—scale, b_4—degree of generalization, i.e.

$$B = \bigcup_{i=1}^{4} b_i.$$

The elements of set C are: c_1—syntactic aspect, c_2—semantic aspect, c_3—aesthetic aspect, c_4—criteria establishing the optimum conditions of perception by man (and for normalized maps—also by computer), i.e.

$$C = \bigcup_{i=1}^{4} c_i.$$

The rules of handling the model, R, include the whole stock of present-day methods of analysing (interpreting) maps, among them primarily cartometry and research methods based on maps.

Cartographic modelling thus includes the processes of developing and using a cartographic model.

A question arises: does the map as a model assure the similarity with the object of modelling, which is necessary for specific investigations? From the similarity theory it follows that in phenomena that are similar in a certain sense one can find certain combinations of parameters, named criteria, having the same values. If phenomena are characterized in space and time with comprehensiveness accessible and necessary for a given investigation, the conditions in this case are regarded as criteria of complete similarity. It is obvious that this condition is best satisfied by large-scale maps, which make it possible to give the most comprehensive description of an object, especially in its geometrical aspect. Scale reduction makes it necessary to generalize the represented phenomena, and that will to a certain extent affect the possibility of preserving their geometrical similarity. It is often difficult, if not impossible, to reproduce a model on the map in full conformity to nature, even in the geometrical aspect (this is equally true for other methods of modelling). Objects that cannot be plotted on the map to scale are shown by non-scale symbols, which exerts an especially negative effect on similarity. With the aim of raising the degree of similarity, the principle of multiscale representation is applied for objects that cannot be mapped to scale (see Chapter 2). By means of this technique, contours which are unplottable to scale are individually represented in other scales, depending on their area. This type of map can be conventionally designated as a similar model, because in it the similarity is revealed by introducing variable scales. It should be noted that a higher degree of similarity is reached in this case, it being possible to establish the correspondence between the parameters of the original and the model. Such similarity can be called quasi-similarity.

When developing a cartographic model it is not always the aim to preserve strictly the geometrical parameters of the geographical data to be represented. A remarkable characteristic of cartographic modelling consists in the possibility of representing the geographical data in a graphic, generalized form to emphasize their most typical, general, and essential features, manifesting themselves to the best advantage in special-purpose generalization. The purpose of the map and its applications are decisive in establishing the methods of the transformation (including generalization) of geographical data in the course of cartographic modelling.

When one models maps to be used in science and industry, where an accurate and detailed representation of geographical data is required, generalization consists in creating a concrete prototype, adequate to the problem in question, by means of formal and non-formal (taking into account the peculiarities of objects) principles of transformation. When one models maps to be used for research purposes, generalization lies in the idealization of the model, predominantly by means of formal transformation principles. This makes it possible, based on the properties of the model, to study the object it substitutes.

Automation of the map-assisted research of objects or phenomena with the help of cybernetic devices requires joint cartographic–cybernetic modelling. Cartographic–cybernetic modelling will be understood as the process of constructing a cartographic model based on a formalized language with its subsequent conversion into a cybernetic model, including the algorithms of data processing.

Let us now examine the generalized scheme of investigating phenomena and processes in the geosphere, based on cartographic–cybernetic modelling (Fig. 1.2). A study of objects in the geosphere (1), usually begins with obtaining the primary data (2) directly by means of field surveys, aerospace photography, collecting statistical data, etc. The process of cartographic modelling itself (3) is carried out proceeding from the primary data. These processes of map compilation are to a certain extent automated. In the automated drawing of normalized maps K_1, K_2, ..., K_n data representation (mainly designing the symbol systems) should be in agreement with the accepted language formalization.

The second process consists in an automatic conversion of a cartographic model into a cybernetic one. This process is controlled by an operator from block 8 using the data visually obtained from block 3, i.e. from the collection of normalized maps. The operator's functions consist in choosing the class of objects or assigning the area for selective automatic readout, input into a computer, and the construction of a cybernetic model.

Block 4 consists of automatic devices P_1, P_2, ..., P_n for selective data readout, capable of identifying the data by their qualitative characteristics. Cybernetic modelling is accomplished by a computer or specialized automatic calculation device (5) operating in accordance with the algorithms and programmes from block 8.

The third process includes a speculative evaluation and analysis (9) of visually perceived data (7) (taking into account the results of research (6), and then synthesis and arrival at preliminary decisions (10) proceeding from the results obtained at the computer output and from the results of analysis performed by the operator.

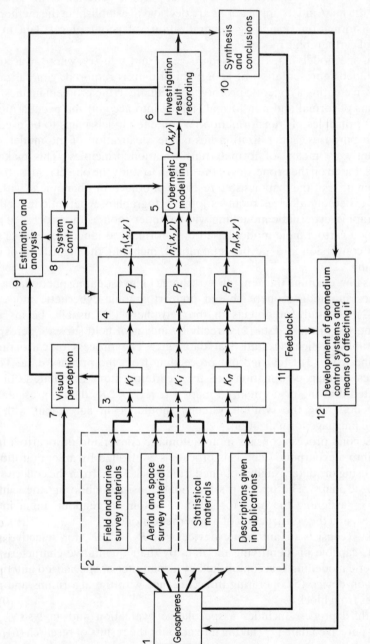

Figure 1.2

Finally, the fourth process presupposes establishing a feedback (11), which includes experimental checking in the field, e.g. in geological investigations—by exploration drilling. It is quite obvious that only after testing the scientific authenticity of the conclusions arrived at will it be possible to develop the methods of control and the means of influencing objects or phenomena in the geosphere. Based on the data obtained (through the feedback channel) and the previously made conclusions and decisions, the systems of controlling the processes and phenomena in the geosphere (12) are developed.

In this way we visualize, in their general form, the functions of an automated system for the investigation of natural and socioeconomic phenomena in the geosphere, where normalized maps serve as the source of initial data to develop a cybernetic model.

In this system the process of developing a cybernetic model incorporates the construction of a mathematical model, i.e. working out the algorithms for the automatic processing of analysed geographical data.

The mathematical model of an object can be developed in different ways: from the results of logical analysis, theoretical research, or previous experience (which is most characteristic for geological and geographical research). The principal directions of research based on maps can be:

1. investigating the interrelations and interdependences of phenomena in the geosphere;
2. forecasting the occurrence of certain phenomena (especially those not readily accessible for direct investigations) by other phenomena;
3. studying the dynamics of natural and socioeconomic phenomena in space and time, etc.

In actual practice, the methods of mathematical statistics have found extensive application.

Cartography today has at its disposal a great variety of methods and techniques to analyse the data presented on maps. Numerous works deal with the mathematical aspects of investigations performed with the help of maps, and only a few of them present the algorithms of computer data processing. There are virtually no works devoted to complete automation (including the automatic readout and input of data into a computer) of investigation processes involving maps. As already noted, the most complicated and as yet unsolved problem is the automatic reading of maps. That is why the input of computer data is for the most part performed manually, which is a very labour- and time-consuming operation. This seriously hinders the application of map-assisted methods of investigation. In this connection it should be noted that the development of an automated processing system based on normalized maps and a formalized language solves this problem.

Cartography, lying close to exact engineering sciences in its objectives, should move more resolutely in this direction so as to equip the branches of science and technology concerned with investigating the Earth and outer space with more refined and accurate means and methods of research.

The described system of investigations performed with the help of normalized maps may become one of the basic systems of geographical data analysis, especially where investigations on vast territories are involved. As to the accuracy and detailedness of normalized map data, they are quite capable of meeting the requirements of numerous research problems, if the chosen scale is appropriate for the objectives of research.

As a model of investigated entity a normalized map has a number of advantages, making it possible:

1. to perform conjugated selective readout and analysis of maps differing in their content, which permits reducing the volume of computer input data and simplifying their processing;
2. to conduct joint computerized and visual analysis of geographical data:
3. to perceive the mutual relations of phenomena.

1.7. Certain Theoretical Aspects of Cartography

The questions of defining cartography as a science, of the substance and objectives of its theory, of its place among other sciences, and its relations with them have been recently attracting the attention of cartographers all over the world. There is no common view on these questions, and A. Robinson [86] is most probably right when he says that it will hardly be reached by the year 2000.

Turning to the history of cartography, one easily sees that the definition of cartography changed with the level of its development, although the definitions cited in literature did not always reflect the most advanced achievements of the science. This is particularly true for the present-day state of the art, and the definitions given by specialists working in different realms of cartography, beginning from the traditional geographic 'cognitive' school and ending with the cybernetic 'formalized communicative' one. The author adheres to a wider understanding of the problems pertaining to cartography: along with informational functions (representation, storage, and communication of data), inherent in cartography are the functions of scientific interpretation of data [120].

Cartography is the science concerned with the methods of representing, storing, and communicating the data on the structure, spatial relations, and properties of terrestrial objects and phenomena and those of celestial bodies, presented in the form of graphomathematical models constructed to an assigned scale (in a generalized and visually perceptible form for man, and a specialized detailed form for computer), as well as with the scientific interpretation of these data for specific practical applications. The 'specialized detailed form for computer' implies images designed only for automatic readout, which can also be presented in the form of latent code.

Some authors, when defining cartography as a science, do not place the functions of data analysis (i.e. their scientific interpretation) among its objectives. Mathematics is known to include, along with purely analytical methods, graphomathematical methods as well; for example nomography, or descriptive geometry. Approaching a map as a graphomathematical model, one is justified in using the cartographic techniques of analysis. Cartography has its own specific language, offering multifarious possibilities. Its basic functions— representation and communication—make it in principle possible to carry out different analytical operations, graphically and graphoanalytically, and strictly mathematically, both by man and with the help of a computer. Cartography in a sense synthesizes certain aspects of such fundamental sciences as mathematics and cybernetics, on the one hand, and geography, geology, geophysics, astronomy, economics, and sociology on the other hand.

The appearance in the 1970s of a new trend—communication in cartography- —exerted a serious progressive influence on the development of the theory. This breakthrough in the development of the communicative trend in cartographic theory was certainly facilitated by the unprecedented scientific progress of recent decades, especially in cybernetics, computer engineering, and computing mathematics. In this connection very much thought has recently been given to cartographic communication in developing the theory. Interest in this aspect of cartography was so keen that it resulted in an unjustified exaggeration of the role of communication in the theory, vagueness of its boundaries, and consequently distortion of the true sense, meaning, and concept of communication in cartography, despite all its positive impact on the development of the theory. One cannot but agree with the critical remarks of J. C. Muller [70] addressed to J. Bertin, to the effect that cartography should not be regarded as a branch of semiology, and a map as a language facility.

Cartography is primarily the science of representing, transforming, and interpreting spatial data with the language of graphics serving as the means of communicating the represented data to the user. The communicative aspect is a derivative of the mathematical aspect, and is represented by signs and symbols performing only one function: to transmit via the visual channel with the maximum clarity (promptly, reliably, vividly) the represented data, the quantitative and qualitative parameters of information (the information content), graphomathematically described in the cartographic form.

In spite of the seemingly narrow limits of its functions, communication is of great importance in the development of the theory. Although many authors in their interpretation of the role of communication in cartography went beyond these limits it did not prevent them from making a sizeable contribution to the development of its theory [7, 9–11, 20, 44, 49–51, 63–65, 67–69, 74, 84, 85, etc.].

The discussion raised in reference [88] seems to be artificial, since within the 'cognitive' aspect of cartography the author K. A. Salishchev has neither devised a theory nor carried out any investigations that could somehow be taken into account in the proposed discussion (with the exception of short definitions, simplistic views on various aspects of cartographic theory, and the critical analysis of literature). Generally speaking, the view on cartography as a science studying the graphical form of representing spatial information does not deny its cognitive role. This is seen from numerous works on communication and automation, including the reference [120], criticized by K. A.

Salishchev, where a number of paragraphs in Chapters I and VII are directly devoted to the questions of investigations performed with the help of maps.

The other trend, automation, is also successfully developing in cartography [8, 12–14, 16–19, 21, 25, 27, 29, 32–35, 38, 41, 42, 56, 57, 59, 61, 73, 79, 80, 82, 87, 89, 120, 135, 142, 143, 146, 148, 150–152, etc.]. These trends exert a mutually stimulating effect upon each other. This is due to their preference for the formalization of science, without which no development of automation in cartography is possible. Despite the obviousness of the major role played by communication in cartography, some cartographers show a clearly negative attitude to this trend. In contradistinction to the formalized communicative trend, in the book quoted in reference [88] the author puts forward a 'cognitive' approach in cartography. Expressing a largely negative attitude to the formalized communicative trend the author at the same time seriously underestimates the role of formalization in scientific theory. Those dealing with the applications of automation and computer technology know very well that without the formalization of theory there can be no progress in science. Theory presents itself in an abstract logical form; it does not reflect the mapped content in concrete manifestations of objects or phenomena, but covers their most general and significant properties which they have in common with other objects and phenomena. Creation and development of a formalized theory are impossible without studying the general common properties, interrelations, and regularities of geographical data. As a result of formalization the theory acquires its most perfect abstract logical form. The higher the level of formalization the more perfect the cartographic model and its language for an objective study of objects and phenomena. Complete automation of the interpretation of a cartographic model is only possible based on a formalized language.

According to Lenin's theory of knowledge, formalization is the refinement of the content of knowledge. Formalization in cartography is a gnosiological device of ascertaining and verifying the most essential and regular aspects of the mapped objects. However, formalization in its broad sense never reaches absolute completeness, and this inevitably leads to a part of the theory remaining unrevealed and non-formalized. This is natural for the development of formal logical means of science and is usually manifested in the appearance of unsolvable problems, which is especially characteristic for such complex processes as automatic identification of cartographic symbols and generalization. For these problems to be eventually solved it is necessary to devise new formal systems, where a part of what has not been previously solved is formalized. Some cartographers believe that formalization and, therefore, automation of generalization processes, do not guarantee ultimate success, as the formalization apparatus and logarithms used cannot account for all the variety and plurality of interrelations of the mapped substantive elements. If generalization follows certain formal principles, i.e. is performed with the help of formalized procedures, it does not necessarily mean that it leaves aside the concrete content, its peculiarities, and the plurality of external and internal interrelations. When a highly developed formalized generalization apparatus is available (providing for interpolation, approximation, filtration, etc.) for concrete objects or their individual characteristics to be represented on the map, appropriate individual procedures are selected, most suitable for their spatial distribution and the purpose of the map.

Formalization in cartography is a complex process requiring from the worker a broad erudition, primarily in mathematics and cybernetics. That is why research and development in this field are so important. It is quite natural that some concrete studies may turn out to be more successful than others. In vigorously developing modern science, formalization and automation inevitably result in the appearance of new terms, concepts, methods, and trends. It is possible that some of them may disappear or prove to be unpracticable, but some will certainly be accepted and developed further.

In world cartography the question of formalizing the science and the language of maps arose for quite natural reasons [1, 51, 53, 104, etc.]. It was a consequence of the general tendency towards formalization of sciences, because without it automation is inconceivable, and automation is one of the determining factors, not only in the development of modern technology, but also in the overall progress of mankind.

The view that communication and formalization play a decisive role in the development of cartographic theory is shared by an overwhelming majority of specialists, as one can judge from numerous publications.

Considering cartographic theory as a certain derivative of substantive geographical analysis, the author in [88] regards the development of the theory as totally dependent on the concrete content. From this view it follows that there can be no theory free of concrete content, there can only be a semi-theory–semi-practice, focused specifically on concrete objectives, constituting in fact a technique tied up with concrete content (geomorphology, soil science, demography, etc.). In other words, geomorphological mapping requires its own theory, geological mapping—its own theory, just as it is the case with demography, etc., which generally speaking is quite justified for the lower level of cartographic theory.

The number of theories will thus be determined by the possible quantity of the mapped thematically connected processes, phenomena, and objects; in other words there are as many theories as there are phenomena. Consequently all attempts to visualize the boundaries of a unified theory are futile, and it is impossible to arrive at a general theory of cartography that would incorporate the most general properties, categories, and principles inherent in a large number of objects and phenomena (geomorphological, tectonic, pedological, etc.) at a more generalized, formally logical level, since in the author's opinion it will be divorced from concrete content.

Thus, proceeding from the views presented in [88] we find ourselves in the province of a narrow pragmatic understanding of theory, which makes it virtually impossible to find any difference between the genuine cartographic theory and practice.

In its theory cartography lies closest to formal sciences (mathematics and cybernetics), and in its practice it lies closest to natural (mainly geological and geographical) sciences. Such a position of cartography explains certain characteristic features in the structure of its scientific theory, where two aspects can be singled out:

1. The formal aspect, associated with mathematicocybernetic methods (interpolation, approximation, filtration, etc.) used to transform (generalize) and represent surfaces and parameters of objects in different systems of coordinates and based on general formal methods of geographical data analysis (applicable to a large number of objects having common properties, interrelations, etc.).
2. The substantive aspect, directly connected with the interpretation of geographical data to identify certain properties, interrelations, and classifications as related to specific problem areas in adjacent fields of knowledge.

The first aspect may be attributed to the upper, abstract, logical level of the theory; and the second to the lower level, less abstract in its form, confined within the boundaries of specific applications of geographical data. The second level is directly connected with practice. It is in the unity of the formal and the substantive aspects that one should see the source of development and improvement of cartographic theory. Representation of substantive elements of the lower-level theory in a formalized communicative form provides for a profound and clear understanding of its structure, logical relationships, etc.

Figure 1.3 shows the logical structure of cartography, where the main emphasis is placed on the theory of science. The structure of the theory is subdivided into two basic levels. The upper level represents cartographic theory proper, with all its parts equally applicable to all geographical data irrespective of their specific purpose. It is thereby the more formalized part of the theory.

The lower level mainly represents the techniques directly associated with the specific applications of geographical data. It covers all the types of specialized (thematic) cartography and has its own methodology of cartographic modelling based on the theory of the upper level. It comprises the methods of generalization, procedures of modelling the legend of maps, techniques of map analysis, etc., but all this as applied to the specific objective (theme) of a given map.

The questions associated with the development of automated systems, as well as information retrieval systems and data banks, are placed at the lower level, since these can only be devised within a specialized branch of science or technology. Computer-oriented formalization is included in all the processes of the theory of modelling and, to a certain extent, the theories of sign systems and of geographical data representation. The technique of choosing the projection as applied to specialized maps is of limited significance (in Fig. 1.3 this link is shown as a dashed line). This technique can be related to, for example, geological mapping of deep-lying structures or astronomical (space) mapping of celestial systems. The general theory of cartographical data representation belongs to the upper level.

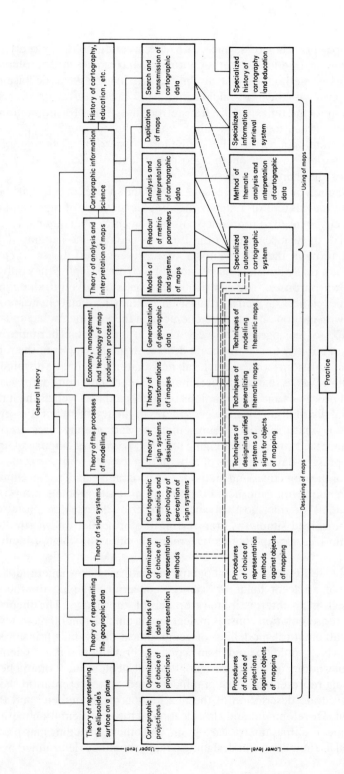

Figure 1.3

The principal branches of science with which cartography is closely connected are physical geodesy, stereophotogrammetry, and topographic and aerospace survey methods. These disciplines provide the geodetic base of a map and supply it with the most valuable initial information. Aerospace survey photographs are sometimes wholly transferred into a map (orthophotomap) and combined with cartographic symbols. Stereophotogrammetry provides a map with accurate data on the relief of the terrain; aerospace photographs—with all the variety of data on the geosphere; and physical geodesy gives information on the shape of the planet.

Any information on any of the geospheres and the spheres of celestial bodies can constitute the content of a map. Cartographic methods of representation are most effectively used to reproduce natural objects or phenomena. It is not accidental that, in such sciences of the Earth as geology and geography, the cartographic form is the leading form of representation. These sciences, however, do not really develop the methodology of cartography, they only use it for their own purposes. The same interrelation is observed in the case of socioeconomic sciences, as well as other sciences studying social phenomena as interrelated with natural objects or phenomena of the Earth.

The method of cartography is very close to those of cybernetics and mathematics. Similarly to cybernetics it deals with the questions of the representation, transmission, storage, retrieval, and analysis of data. Its distinguishing feature, however, is that the subject matter of representation comprises the most general processes and interrelations of phenomena, as well as the nature of their spatial distribution over the geospheres of the Earth. What is essential in this case is that, for data representation, cartography uses its own graphomathematical language belonging to a special group of languages differing from the present-day languages of cybernetics.

Similarity of the objectives and methods of cartographic data representation and transformation with cybernetics cannot but pose the problem of utilizing the achievements of this most advanced science. The problem, however, should not be solved simply by transferring the principles and methods of cybernetics into cartography, but by creatively adopting them to suit its objectives.

Cartographic representation of geographical data has its own methodology, independent of adjacent fields of sciences concerning the Earth where it manifests itself as an interdisciplinary branch of knowledge. The theory of cartographic representation consists in developing unified general methods for the graphomathematical modelling of geographical data, their transmission and storage, which could be used in any concrete branch of adjacent sciences, irrespective of whether astronomical, geographical, geological, or other phenomena are being represented. Cartographic methods of representation should provide an accurate description of the distribution of phenomena and their interpretation. Development and classification of the representation methods (isoline method, multiparametric cartogram, and others, including map projections), as well as their choice in designing a map, should be determined by the nature of the distribution of objects or phenomena (continuous or discrete), by a combination of qualitative and quantitative parameters and the metric of representation employed, and by the contemplated use of the map.

Cartographic methods of representation are applicable to any geographical data, the same methods often being equally suitable for data with a different content. For instance, the isoline methods can be used with equal success to represent the relief, magnetic fields, precipitation, and any other data of similar distribution. The same can be said of the conformal cylindrical or any other projection that can be used for the representation of population, vegetation cover, geophysical fields, and other objects and phenomena distributed over the geosphere.

The methodology of cartographic representation depends on neither the concrete content nor the practical applied problems. Its objective is to develop unified principles, unified methods of representing geographical data; mainly in their more general, abstract forms of existence. Its graphomathematical methods of representation should make it possible to describe both the spatial distribution of the individual objects or phenomena, and the general nature of spatial distribution characteristic for a large number of objects and phenomena. In other words, its methods must be able to represent any spatial objects and phenomena, irrespective of whether there is information on their concrete qualitative differences, but which are known to have a common and, at the same time, specific nature of distribution on the geosphere, including their inner spatial interrelations, interactions, etc. This, however, should not be understood to deny the necessity of taking into account the concrete content of the designed map, of qualitative interpretation of the objects to be mapped when one is designing a concrete map. The relationship between the methodology of cartography and the practical mapping is the same as that between theory and practice in other branches of science.

Development of cartographic theory should be focused on ascertaining the general nature of the distribution of objects and phenomena in geodetic and free space, and identifying their similar properties by studying their concrete individual manifestations so as to develop appropriate representation and interpretation methods based on graphomathematical modelling. The trend in question conforms in a natural way to the dialectical method of cognition—from the individual to the universal. And practical cartography should be directed at using these methods with due account for characteristic individual features of the mapped objects and phenomena, i.e., by means of representing the individual and specific through the common and universal. At the same time, the methodology of cartography must embody representations and interpretations of both the specific, individual, single and the common and universal, which can be manifested in all processes and objects of the geosphere, in principle associated with all the sciences concerning the Earth and, to a certain extent, with socioeconomic sciences. It is here that we see the independence of cartography from the concrete sciences of the Earth.

The independence of cartography is also easily proved by the fact that cartographic methods can be applied, for example, to represent and study the spatial distribution of micro-organisms in biophysics, of microstructures in flaw detection, of fields in physics, of celestial bodies in space, and many other data having nothing to do with the adjacent sciences of the Earth.

CHAPTER 2
New Cartographic Methods for the Representation of Geographical Data

This chapter deals with new methods of representing geographical information. A considerable part of the chapter is devoted to the method of raster discretization (MRD).

MRD has been developed taking into account the logical potential and the resolving power of computers, as well as human psychology and man's ability to read and perceive the cartographic image. The method makes it possible to represent objects or phenomena characterized by practically any spatial distribution pattern, with the exception of linear objects.

MRD, as distinct from the traditional methods of representing information, permits showing within an enclosed area a number of qualitative and quantitative features of objects. This property of MRD makes it quite an efficient tool for depicting multiparametric objects.

A comprehensive analysis of MRD has shown that it can be used to represent all kinds of information, both continuous and discrete, irrespective of the nature of its spatial distribution, though it cannot be equally effective in all cases. Thus, for example, to depict certain classes of information represented by a continuous $z(x, y)$ function, the method of isolines is more effective in a number of cases.

Therefore the representation technique is examined in this chapter as applied to such objects or phenomena for which MRD is the most expedient. In general, MRD can be primarily expedient where it is necessary to represent information of an analytical nature, distinguished by a high degree of detailedness and accuracy, to be subsequently utilized in scientific research.

The techniques of representing linear (extensive) objects with the help of the multiscale (over the area and length) method are also described in this chapter.

2.1. The Method of Raster Discretization

The principal objective of the method is to universalize the graphical means of formalized representation of information, based on graphomathematical representation techniques. The purpose of MRD is to provide for the representation of any kind of spatial geographical information, both discrete and continuous, and to assure that a map of any content plotted with the help of MRD could be selectively read out with one and the same complex of software for specialized input devices and a computer. A major requirement is to ensure a high degree of detailedness and accuracy in representing information, with sufficiently good visual readability of the image.

The gist of the method consists in discrete graphomathematical representation of information, based on a system of linear rasters of differing parameters, which at the same time serve as informative and illustrative features of the qualitative and quantitative aspects of the map content. Raster lines are taken with the same spatial orientation, i.e. the same direction (in some individual cases a mutually perpendicular arrangement of raster lines is permissible).

The cartographic information represented by MRD is read out by photo-electron scanning devices with the scanning lines going perpendicular to the raster lines. The informational parameters of such a raster system are:

1. Thickness T, characterized by: the difference between raster line thicknesses, when read visually (Fig. 2.1b); the time interval τ determined by the width of the videopulse appearing at the moment when the scanning element traverses the raster line, when read by a computer (Fig. 2.1a).

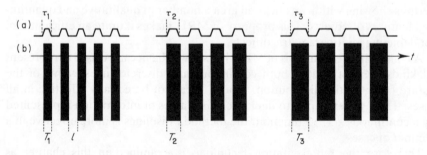

Figure 2.1

2. Density W, characterized by: the interval, l, between the raster line axes, when read visually (Fig. 2.2b); the recurrence frequency of videopulses, $n/\Delta t$, formed by the raster lattice in the course of scanning, when read by a computer (Fig. 2.2a).

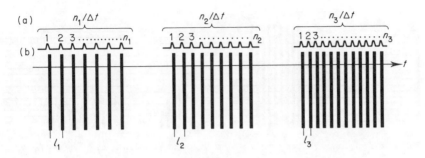

Figure 2.2

3. Colour C, characterized by: the hue, colour saturation, and brightness variations, when read visually; the videopulse amplitude, U, and the optical spectrum zone, $\Delta\lambda$, when read by a computer (Fig. 2.3a).

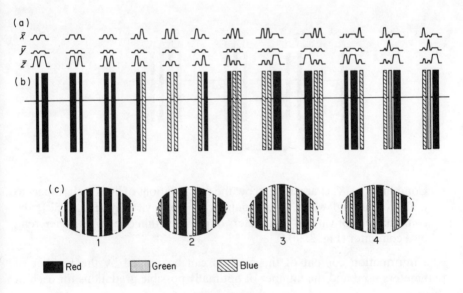

Red   Green   Blue

Figure 2.3

4. Length L, characterized by: the difference in the length of individual lines or by the total length of raster lines, when read visually; the line advance, d, of the scanner and the number of pulses, N, appearing at the moment when the scanning element traverses the raster lines, when read by a computer (Fig. 2.4).

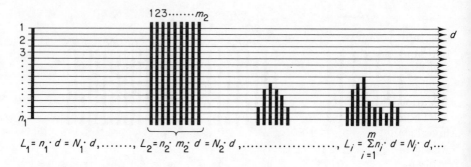

$$L_1 = n_1 \cdot d = N_1 \cdot d, \ldots\ldots, \quad L_2 = n_2 \cdot m_2 \cdot d = N_2 \cdot d, \ldots\ldots\ldots\ldots\ldots, \quad L_j = \sum_{i=1}^{m} n_i \cdot d = N_j \cdot d, \ldots$$

Figure 2.4

5. Groups H, characterized by: the number of lines in a group (series), m, with equal intervals between them, e.g. consisting of one, two, three and more lines, when read visually; the number of pulses in a series having an equal frequency, when read by a computer (Fig. 2.5).

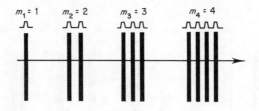

Figure 2.5

6. Combinations K, characterized by: the combinations of lines with respect to their colour and width in a group, when read visually (Fig. 2.3b); the combinations of various parameters of pulses within one series, when read by a computer (Fig. 2.3c).

The information content of the method can be judged by the number of parameters used and the number of optimally possible gradations for each of them.

The information content of the method will depend on the following two principal factors:

1. optimal discrimination thresholds with respect to each parameter (i.e. the number of gradations) for man and computer;
2. requirements on detailedness and accuracy of the represented information (note that an increase in the number of gradations for parameters T, W, and H in some cases reduces the detailedness and accuracy).

2.2. The Informational Properties of MRD

The information content of the method is associated with the problem of the optimum choice of raster parameters. The information content of MRD will be considered in this book to be the quantity of raster elements variety, i.e. the number of different parameters and the combinations of raster line parameters.

Variety is understood as the difference between the elements of a set. For MRD one has to consider the difference in the parameters of individual raster lines having reliable discrimination thresholds for both computer and man, as well as the differences in the combinations of these parameters.

Let us now define the criteria on which one can judge the information content of MRD. Since the factor that determines the information content is the number of discrete decomposition elements (gradations) of each parameter in the raster system, we shall examine them individually.

The main difficulty in determining the criteria of informativeness lies in the necessity of simultaneously taking into account the capabilities of man and computer. While visual discrimination thresholds for man are to a certain extent already known, for a machine (in this way we shall briefly designate any cybernetic reading device) the analysis of its resolving power presents a considerable difficulty. This difficulty is associated with the indeterminacy of the machine class. Investigations of different devices for the conversion of graphical information have shown that, generally speaking, the resolving power of present-day cybernetic devices (particularly those equipped with scanners having microscope objectives) exceeds the resolving power of a human visual analyser.

In this connection, when determining the number of possible gradations for all the parameters we shall mainly proceed from the capabilities of the visual analyser. Let us consider the criteria used to determine the number of gradations for line thickness.

The number of possible gradations within the line thicknesses is a function of the minimum line thickness $T_{\min}$, its maximum thickness $T_{\max}$, and the optimal coefficient of line thicknesses discrimination ρ at brightness contrast r_i

$$n_T = f(T_{\min}, T_{\max}, \rho_{T,r_i}) \tag{2.1}$$

The optimal discrimination coefficient must provide for a sufficiently reliable identification of lines by their thickness on the map field at the minimum possible discrimination threshold. Brightness contrast is determined from the ratio of line brightness, B_i, to background brightness, B_0.

For rasters with the brightness contrast $r_i = 0.14$ (black lines against the white background of paper) the discrimination coefficient $\rho_{T,r_i} = 1.55$. The number of raster gradations within line thickness can be calculated from formula

$$N_n = N_1 \times \rho_{T,r_i}^{n-1} \tag{2.2}$$

where N_1 stands for the initial parameter of visual stimulus; in this case the minimum line thickness, T_{min}.

Thus, for example, if $T_{min} = 0.1$ mm and $n_T = 6$ we shall get a scale with the following values: 0.1; 0.15; 0.25; 0.40; 0.60; 0.95 (after a slight rounding off of the second decimal digit). But, taking into account the editing of maps, the scale to be recommended is: 0.1; 0.2; 0.35; 0.55; 0.8; 1.2.

In designing the legend for a map it is advisable to confine oneself to a limited number of colours within one symbol. It is therefore important to calculate the number of combinations with a preassigned number of elements in a group. If a set contains n elements, its subsets consisting of k elements are designated as a combination of n elements taken k at a time, or, in our case, as a combination of n colours used on the map taken k colours at a time in a group. Their number is designated as N_n^k. Proceeding from the total number of subsets being equal to 2^n and applying the elementary principles of combinatorics we can write the equality:

$$N_n^0 + N_n^1 + \cdots + N_n^k + \cdots + N_n^n = 2^n \tag{2.3}$$

The value of N_n^k can be calculated from formula:

$$N_n^k = \frac{n!}{k!(n - k)!} \tag{2.4}$$

where $k! = 1 \times 2 \times 3 \times \cdots \times k$ and $O! = 1$.

Using the formula one can calculate the number of combinations (in colours or line thicknesses) without taking into account the permutations of lines in a group.

Let us consider the information content of the method in terms of the number of combinations resulting from all possible variations of the thickness and colour of lines in a group.

We shall begin with the principle of calculating and constructing the parameters of lines formed without taking into account the specific order of combinations, i.e. without taking into account the possibilities of their permutations inside the groups, and then the principle for the case of various transmutations of the colours and thicknesses of raster lines in the group. It should be noted that completely free permutations of the colours and thicknesses of raster lines are only possible in groups the interval between which is no less than twice the regular interval taken between the lines in the group, otherwise the combinations become unidentifiable.

Following the first principle we can also present all the lines with a constant interval, since here the difference between combinations will be identified simply by the presence of fixed colour and thickness parameters in the group, irrespective of their sequence (arrangement). The first principle, producing combinations with a uniform interval between lines, as distinguished from the second, makes it possible to construct symbols with a uniform raster background, but the information content in this case will be considerably lower.

One can obtain the total number of combinations from formula

$$N = n_{T_0}^{n_c} - 1 \tag{2.5}$$

where n_{T_0} is the number of line thicknesses together with the null thickness; n_c is the number of colours used in a group. Note that in this case the maximum number of lines with a different colour in a group is equal to the number of colours used.

Taking into account the number of groups n_H (if they have double intervals between them), the number of combinations can be calculated from formula

$$N_k = (n_T^{n_i} - 1) \times n_H \tag{2.6}$$

The variant with free permutations within the groups has the maximum information content.

The number of possible combinations for it can be calculated from the formula:

$$N_k = (n_c \times n_T)^m \tag{2.7}$$

where m is the number of lines in a group.

It is seen from the formula that the groups make it possible to represent hundreds and even thousands of qualitatively different objects with quantitative characteristics. Thus, e.g., in a 'ternary' system, when $n_c = 3$, $n_T = 3$, and $m = 3$, 729 different characteristics can be shown on a map. With the data processed by a machine the possible number of combinations is only limited by its storage capacity (when reference patterns for identification are designed). A machine can identify any of the above combinations.

2.3. Representational Properties of MRD Rasters and their Perceptibility for Man and Machine

Representational properties of MRD rasters will be considered as compared with the traditional cartographic techniques.

Generally speaking, rasters (or grids of different configurations) as a representational means have been known in cartography for a long time.

In terms of representation the principal distinction of MRD rasters is the uniform direction of all the raster lines at any combination of their parameters. This restricts the representational properties of MRD as compared with an arbitrary raster but, at the same time, MRD introduces orderliness and logic into the system of designed symbols (which is especially important for machine readout) ensuring a high information content. This is achieved by the lines of the raster being plotted in parallel when different parameters are combined, owing to which they are perceived individually in any combination. This unified representation style makes it easier to introduce the necessary metric, and allows a more accurate comparison and estimation of information, both by measuring devices and visually.

Unlike the traditional cartographic language, MRD constitutes a unified graphically and algorithmically formalized cartographic language. MRD consists of fundamentally new representation techniques, such as the technique of a multiparametric continuous cartogram, the technique of digital code symbols, or the multiscale representation technique. From the point of view of representation, MRD enables one to obtain a uniform colour field by utilizing the effect of the spatial mixing of colours. The effect of spatial mixing of colours is known to be observed when the surface of paper covered by coloured lines is perceived at a certain distance as uniformly monochromatic. A fully smoothed-out and uniform tint is obtained when lines are imperceptible. However, the unity of colour in perceiving parallel-plotted lines of different colour is also observed when the lines are still quite discernible. This is a result of the regular repetition of monochromatic lines over the field. The effect of perception duality lies at the basis of the representational properties of MRD. It should be noted that MRD permits the existence of a weak background filling on which the raster lines can be then superimposed. Such a combination increases the effect of colour field uniformity. Maps printed in such a manner outwardly do not differ at all from traditional maps.

Depending on the representation technique, certain raster line parameters (thickness, density, colour), do not necessarily have to be perceived visually. This is, for example, the case with some maps displaying a qualitative background. It is only necessary for a machine to be able to identify these parameters. Parameters of lines may be so small in size that they cannot be visually perceived; that is, they lie beyond the visual perception threshold. Here the ability of the human eye to integrate, and generalize the image, comes into effect, resulting in spatial colour mixing. It is known that the thinner and more dispersed are the lines filling a contour, the lighter is the perceived colour of the background and in contrast the thicker and closer to each other are the lines, the darker is the perceived background within the contour. Something like that also occurs with colour mixing. If lines of different colours (e.g. red and blue) alternate within a contour the contour will be perceived in the integral colour—in this case, lilac. The colour can be varied from purple to blue–violet by changing the thickness of one of the lines. A similar effect can be obtained with lines of the same thickness, by varying their number in a group. Experiments on composing various kinds of analytical and synthetic maps, with the information on them represented by MRD in conjunction with other techniques, show that in their outward appearance these maps do not differ at all from traditional ones. Another advantage of MRD is that the uniformity of raster line directions assures their reliable identification by a machine and the measurement of the quantity of information they carry. All this is done in a unidimensional space (unlike the traditional images requiring a two-dimensional analysis). This is achieved owing to the scanning element intersecting the raster lines perpendicular to their direction. The readout of images is performed on serially produced scanning devices.

The possibilities of image perception offered by MRD to man and machine, in comparison with traditional images, will now be examined as exemplified by discrete data represented by means of MRD and using the dot technique. The latter is known to be widely used for traditional maps, where it is mainly applied to represent dispersed objects.

Let us now consider the perceptibility of a machine with respect to the dotted and the linear raster images.

The readout of such images is most expediently performed with the help of linear scanning. It is in principle possible to calculate the number of dots and their size, and then to determine the quantitative value for an assigned area. In order to obtain the data for a dot (i.e. its parameters), needed for computer processing, it is necessary to scan by lines (as a minimum, two lines per dot) and then to analyse the obtained data in a two-dimensional space. Even with quite closely spaced scanning lines one can hardly expect to arrive at sufficiently accurate dot diameter values.

It is much simpler to read and process the information represented by MRD linear raster. Owing to the scanning lines being perpendicular to the raster lines it is possible to determine the line thickness with reasonable accuracy. Displacement of scanning lines in either direction can only cause an error along the line, which will not be greater than one scanning step (line advance). The reading of images can be performed on simple scanning devices.

Finally, as already mentioned, line thickness determination—and, in general, symbol identification by MRD—can be achieved with single-line scanning, whereas traditional images require multiple scanning.

In conclusion it should be noted that MRD assures a higher perceptibility for machine and man in the readout of quantitative parameters than do the traditional representation techniques (if the information content of the compared images is the same). Moreover, a linear raster is also simpler for the plotting of cartographic images with the help of graph plotters and computers. At the same time the qualitative characteristics of traditional images are visually perceived with higher reliability, which is associated with the freedom of choosing the graphical patterns in designing the traditional symbols.

2.4. Techniques of Representing Discrete Information

Depending on the purpose of the map, discrete information can be represented in two ways:

1. over the area of occurrence within the limits of administrative or natural boundaries (characterizing its density, qualitative features, etc.);
2. locally at the points of its maximum concentration (characterizing its density, qualitative features, etc.).

For the first representation mode the following techniques are recommended:

(a) multiparametric continuous cartogram (MCC) prepared by MRD;

(b) continuous proportional cartogram prepared by MRD;

(c) diagrammatic map prepared by MRD.

Each of these techniques has its advantages and disadvantages. The choice of technique is closely associated with the nature of the data at the cartographer's disposal. If, for example, the data on population are only available for administrative units (territories), there is no other choice but to apply either the cartogram or the diagrammatic map technique. If the data on population are available for populated localities any of the above techniques can be applied. Essential in this case is the possibility of showing the characteristics of the density and other population distribution parameters within their natural boundaries; e.g. with the help of a multiparametric continuous cartogram.

Of the above-mentioned techniques, that of multiparametric continuous cartogram is the most accurate and informative. It also produces a more objective graphic representation and is the most adequate for maps having scientific and industrial applications.

For the second representation mode the following techniques are recommended:

(a) the method of digital code signs prepared by MRD;

(b) the method of raster symbols prepared by MRD.

The technique of digital code signs prepared by MRD provides better accuracy and detailedness.

2.4.1. Representation with the help of multiparametric continuous cartogram technique

The technique in question makes it possible to simultaneously represent within the boundaries of given territorial units the density and absolute number of objects, their qualitative characteristics, and percentage ratios [105, 121]. In individual cases, if some additional characteristics have to be introduced, perpendicular lines may be used. One should, however, be cautious in this case so as not to detrimentally affect the readability of the map.

It is noteworthy that a map composed in this way facilitates taking accurate data from it using a meter and a rule, as well as automatic devices, with high visual clarity being preserved.

The initial data on which maps are composed by means of the MRD multiparametric continuous cartogram technique are: statistical data, coordinates of the boundaries of territories and their areas, and graphical data for the boundaries of an area to be mapped.

Let us now consider the procedure of designing maps which characterize the density of discrete objects within territorial units.

Depending on the magnitude of density variations within the mapped area there can be two variant solutions. We shall first consider the case of minor variations in the density of discrete objects. Here the use of only the raster line density (at an equal thickness of all the lines) may prove to be sufficient.

The density of raster lines can assume different values, depending on the number of discrete objects, the area of the territory they belong to (i.e. within which they are presented), and the weight of a unit raster line length. The frequency of lines is calculated from the following formula:

$$l = \frac{S \times p}{0.01 \times N \times M^2} \tag{2.8}$$

where l is the distance between the raster line axes, mm; S is the area of the territory, km^2; p is the weight, representing the number of objects per unit of length (e.g. 1 mm or one scanning step); M is the denominator of the scale of the plotted map, km; N is the number of objects.

It is obvious that the smaller the value of l the higher the density of objects, and vice-versa. When choosing the weight p it is necessary to proceed from the optimum variant, i.e. in the area with the maximum density of objects the raster lines should have the maximum admissible density, and when the density of objects is small they should not be excessively rarefied.

If density variations within the mapped region are so substantial that in low-density areas the raster becomes excessively rarefied and in high-density areas it becomes excessively dense (especially when the lines begin to merge), it is expedient to use several weights, $p_1, p_2, \ldots, p_n$, differentiated by the thickness or the brightness contrast of lines (when the objects are qualitatively alike, differentiation by the colour hue is possible).

Practical experience has shown that in most cases it is sufficient to introduce two or three different weights and, correspondingly, two or three lines of different thickness to designate them.

The technique used for representation can differ depending on the form of the initial data. Let us examine the specific variants of the representation technique.

1. *Initial data are presented as limited interconnected regions having no great differences in the area and the density of discrete objects (e.g. population of administrative districts, provinces)*

In designing a map it is most important to determine the optimum weights, $p_1, \ldots, p_n$. The first to be found are p_{max} and p_{min}. The following formulae are used for this:

$$\begin{aligned} p_{max} &= 0.01 \times l_{min}^{(T_{max})} \times p_{max} \times M^2 \\ p_{min} &= 0.01 \times l_{max}^{(T_{min})} \times p_{min} \times M^2 \end{aligned} \tag{2.9}$$

where $p = N/S$ (i.e. the density of discrete objects per 1 km); $l_{min}^{(T_{max})}$ is the minimum possible distance between the raster axial lines at their maximum thickness (or brightness contrast) with due account for the gradations whose number is equal to that of the taken weights; and $l_{max}^{(T_{min})}$ is the maximum possible distance between the raster axial lines at their minimum thickness T_{min}.

Parameters T_{min} and T_{max} are calculated in accordance with the number of gradations. The value of $l_{max}^{(T_{min})}$ is chosen so as to meet the condition of

$$l_{max}^{(T_{min})} \leqslant \frac{L_y}{3} \tag{2.10}$$

where L_y is the smallest extent of the areas, measured along the axis perpendicular to the direction of raster lines. In other words, it is necessary that no less than two lines should lie within the area of minimum extensiveness. The value of $l_{min}^{(T_{max})}$ is calculated from the formula $l_{min}^{(T_{max})} = e + T_{max}$, where e is the interval between lines. This value is determined taking into account the optimum visual discrimination threshold, the performance characteristics of polygraphic equipment and reading devices (it is of the order of 0.1–0.2 mm).

If the differences in weights are given not in terms of line thickness, but of brightness contrast or colour hue, parameter T then remains constant, i.e. $T_{min} = T_{max}$. In this case representation will be more accurate, since one can take the minimal value for $T = $ const., which means that the distance between axial lines will be shorter, and therefore the weight will be smaller too. Obviously, the smaller the weight the more accurate the representation. The number of weights can be judged by the difference between p_{max} and p_{min}. If $p_{max} \leqslant p_{min}$, p_{max} can be taken as the only weight. If $p_{max} > p_{min}$, a minimum of two weights is taken. The greater the difference between p_{max} and p_{min} the greater the number of weights. The number of weights taken will also depend on the purpose, scale, accuracy and detailedness of the designed map, as well as the nature of spatial distribution of objects on it.

When composing a map the distribution of weights with equal intervals between their values can be performed in accordance with the following simple algorithm:

1. the value of the interval is found:

$$\Delta p = \frac{p_{max} - p_{min}}{n - 1} \tag{2.11}$$

   where n is the number of weights;
2. the scale of weights is constructed:

$$\left. \begin{aligned} p_1 &= p_{min}, \\ p_2 &= p_1 + \Delta p, \\ &\cdots\cdots \\ p_j &= p_{j-1} + p, \\ &\cdots\cdots \\ p_n &= p_{max}; \end{aligned} \right\} \tag{2.12}$$

   the weights in this case are rounded off to the nearest higher value.
3. the regions are distributed according to weights, matching the intervals of differing densities, Δp

$$
\left.
\begin{array}{ll}
P_{\min} \div P_{\min} + \Delta P & \text{for the weight } p_1 \\
P_{\min} + \Delta P \div P_{\min} + 2\Delta P & \text{for the weight } p_2 \\
\cdots\cdots\cdots\cdots\cdots\cdots\cdots \\
P_{\min} + (n - 1)\Delta P \div P_{\max} & \text{for the weight } p_n
\end{array}
\right\} \quad (2.13)
$$

where $\Delta P((P_{\max} - P_{\min})/n)$.

2. *The regions have small differences in area but differ greatly in the densities of discrete objects*

$P_{\min}$ and $P_{\max}$ are determined as in the first example, and a scale is constructed with subdivision into gradations (intervals). To assure good readability of images it is desirable to have not more than five intervals.

If $P_{\min}$ is a fraction it is expedient to construct a scale with diversified intervals, starting with small ones from $P_{\min}$ and increasing up to $P_{\max}$. The scale can be logarithmic. If the value of $P_{\min}$ is an integer, the intervals of the scale can be equal. The value of $l_{\max}^{(T_{\min})}$ is found from formula (2.10) for the region the least extended along the chosen coordinate axis within a given interval. Weight p_j for the jth interval is assigned proceeding from the following conditions:

(a) if $p_{\min}(j) > p_{\max}(j)$ then $p_j = p_{\max}(j)$ or $1/2p_{\min}(j) + p_{\max}(j)$;
(b) if $p_{\min}(j) < p_{\max}(j)$ then $p_j = p_{\min}(j)$,

but in this case the upper boundary of density, P_j, for the jth interval and, therefore, the whole scale, will change. The value of P_j is calculated from formula:

$$
P_j = \frac{p_1}{0.01 \times M^2 \times l_{\min}^{(T_j)}} \quad (2.14)
$$

In accordance with the found P_j value all the subsequent series of interval value also changed.

3. *The regions have large differences in area and in the densities of objects*

The scale of intervals is assigned similarly to the previous variant, the weights are calculated and the intervals determined as shown in formula (2.15).

The intervals are calculated until the condition of $P_j \geqslant P_{\max}$ is reached. The obtained values of weights, p_j, and densities, P_j, are rounded off to the nearest lower value. The aim is to obtain a scale which would represent in the best possible way the occurrence of the phenomenon in question for the visual evaluation of the map.

1. $p_1 = 0.01 \times M^2 \times l_{max}^{(T_1)} \times P_{min}$

$$P_1 = \frac{p_1}{0.01 \times M^2 \times l_{min}^{(T_1)}} \qquad \qquad P_{min} \div P_1 \text{ for the weight } p_1$$

2. $p_2 = 0.01 \times M^2 \times l_{max}^{(T_2)} \times P_1$

$$P_2 = \frac{p_2}{0.01 \times M^2 \times l_{min}^{(T_2)}} \qquad \qquad P_1 \div P_2 \text{ for the weight } p_2$$

3. $p_3 = 0.01 \times M^2 \times l_{max}^{(T_3)} \times P_2$

$$P_3 = \frac{p_3}{0.01 \times M^2 \times l_{min}^{(T_3)}} \qquad \qquad P_2 \div P_3 \text{ for the weight } p_3$$

$\qquad \left. \begin{array}{c} \\ \\ \\ \\ \\ \\ \\ \\ \\ \\ \\ \\ \end{array} \right\}$ (2.15)

$\cdot \quad \cdot \quad \cdot \quad \cdot \quad \cdot \quad \cdot \quad \cdot \quad \cdot \quad \cdot \quad \cdot \quad \cdot \quad \cdot \quad \cdot \quad \cdot \quad \cdot \quad \cdot \quad \cdot \quad \cdot \quad \cdot \quad \cdot$

4. $p_j = 0.01 \times M^2 \times l_{max}^{(T_j)} \times P_{j-1}$

$$P_j = \frac{p_j}{0.01 \times M^2 \times l_{min}^{(T_j)}} \qquad \qquad P_{j-1} \div P_j \text{ for the weight } p_j$$

After finding the weight values the frequency of lines is determined for each territorial unit from formula (2.8). Taking into consideration that the $p_j/(0.01 \times M^2)$ ratio in this formula is constant for all the regions whose weight is p_j, we shall designate

$$\frac{p_j}{0.01 \times M^2} = \theta_j \qquad (2.16)$$

and since $P = N/S$ the formula will assume the following form:

$$l_{ij} = \theta_j \frac{1}{P_i} \qquad (2.17)$$

When one weight is given, the parameter of line thickness can be used to represent some other characteristic. It is expedient to represent qualitative characteristics of objects by colour.

For simplicity we shall henceforth refer to quantitative values of various objects as indices.

The indices of objects can be expressed in percentages. For instance, having taken one line filling a contour as accounting for 10 per cent of a group consisting of 10 lines, the percentage can be expressed by the number of lines of the same colour in the group. If the required accuracy is higher than 10 per cent the units of percentage can be shown as proportional segments along the length of the lines. In a simplified version one line can be taken as accounting for 20 per cent; the group will then consist of five lines. It is recommended that the lines of one index in a group should be compactly spaced, i.e. lie next to one another.

It should be noted that, irrespective of the number of qualitative characteristics, the frequency of lines is assigned in accordance with the total density of all the indices.

From such maps the following data can be obtained:

1. the overall density of objects for separate regions or the average for a group of regions;
2. the density of objects in terms of the individual qualitative indices making up the complexes;
3. the percentage of indices;
4. the overall quantity of objects for separate regions and for a group of regions;
5. the quantity of objects in terms of indices;
6. the areas of each region and a group of regions.

Concrete data can easily be obtained for each of the above-mentioned characteristics with the help of simple instruments (a rule and dividers), as well as by automatic means. Let us now examine the possibilities of taking from the map the data for each of the characteristics.

1. The value of the density of objects, P_i, can be obtained from formula (2.14).

As seen from the formula, one measurement of the distance between axial lines of the raster is sufficient to determine the density. For convenience it is recommended to prepare a special nomograph, with the density of objects per 1 km plotted on the horizontal axis and divisions in mm on the vertical axis. Figure 2.6a shows a schematic diagram of the population of Kazakhstan, and the corresponding nomograph is shown in Fig. 2.6b. Each weight value, p_j, has its own curve. The curves are calculated from formula (2.17).

To raise the accuracy of density data readout from the map with the help of dividers and a rule (especially for such places where the density is high and the lines are very close) it is additionally recommended to give curves with the weights increased several times, e.g. with weights $5p_1, 5p_2, 5p_3$, etc. The upper part of Fig. 2.6b shows the auxiliary p'_1, p'_2, p'_3 curves with a five-fold increase in weight. Then, evidently, in those places where the lines lie very close to one another the density should be measured not between the adjacent lines but over intervals.

For the convenience of visual estimation of data it is recommended that a scale should be plotted, where the intervals of density values are given in accordance with the frequency and the thicknesses of lines (Fig. 2.6c).

56

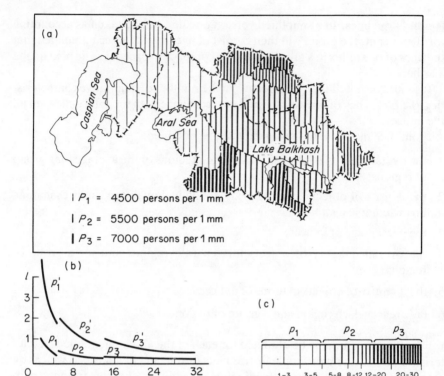

Figure 2.6

2. The density of objects in terms of their indices is also determined with the help of a nomograph. If the density of a concrete index is represented by a line alternating with others, i.e. inside a group, the distance is measured between the lines of the same colour ascribed to the given index. This distance is then plotted on the nomograph, and the distance between the corresponding curve and the horizontal axis yields the density value. If the density of the given index is represented by several lines running in succession in a group, then the interval is measured between the first lines of the same colour, which denote the index, in two adjacent groups. The value read from the nomograph is multiplied by the number of lines in the group that have the same colour, i.e. denote the given index. The result obtained will characterize, with a certain approximation, the density of this index. In the absence of the nomograph the density of objects in terms of their indices, P_{ij}, can be calculated from formula:

$$P_{ij} = \frac{p_{ij} \times \varepsilon}{0.01 \times l_i^1 \times M^2} \qquad (2.18)$$

where ε is the number of lines denoting the given index in a group; l_i^1 is the interval between the first lines of the same colour in two adjacent groups.

3. The percentage of objects in terms of indices is calculated from formula

$$N_\% = \frac{\varepsilon}{F} \times 100 \qquad (2.19)$$

where $N_\%$ is the percentage of the given indices in all the indices for the region; F is the total number of lines in a group.

4. The number of objects in a region is determined as the total length of all lines,

$$\sum_{i=1}^{n} L_i$$

falling within the limits of a region, multiplied by weight p_j. The total number of objects, N_{tot}, for m regions will be expressed by the formula:

$$N_{\text{tot}} = p_j \sum_{k=1}^{m} \sum_{i=1}^{n} L_{k,i} \qquad (2.20)$$

where k is the number of the region and i is the number of the line.

5. The number of objects characterized by the same index is determined in a way similar to that for the total number of objects, with the difference that only the lines denoting the index in question are taken into account.

6. The area of a region is calculated from formula

$$S_k = \sum_{i=1}^{n} L_i \times l_k \times M^2 \qquad (2.21)$$

The accuracy of determining the number of objects and the area from the total length of lines,

$$\sum_{k,i=1}^{n} L_{k,i}$$

with due account for the weight, p_j, and the interval between lines, l, respectively, depends on line spacing, contour configuration, and weight p_j. The closer are the raster lines to each other, and the smoother the region boundary, the more accurate the representation. Representation error can be of either sign. Because of the fact that line spacing is a variable quantity, i.e. it can change within considerable limits, and also that several weights can be used in representation, the values of errors for different regions can be quite different. The studies of accuracy reported in reference 157 show that with a decrease in line density (increase of l) the error, V_l, increases linearly (Fig. 2.7), and with an increase in area S, at constant line spacing, the error decreases in a nonlinear fashion (Fig. 2.8).

Taking into account the fact that at small areas and wide intervals between raster lines gross errors are possible, one has, when using this representation technique, to introduce corrections for the lengths of lines filling the contour.

58

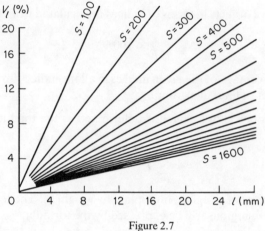

Figure 2.7

In actual practice this is done by means of simple calculations, based on determining the error, δ, as the difference between the measured value (found from formula

$$N^1_{k,i} = \sum_{k,i=1}^{n} L_{k,i} \times p_j$$

within the limits of the kth contour) and the initial value, $N^1_k - N_k = \delta_k$. If the error has the positive sign, the total length of lines,

$$\sum_{k,i=1}^{n} L_{k,i}$$

is reduced by the value of

$$\Delta L_k = \frac{\delta_k}{p_j}$$

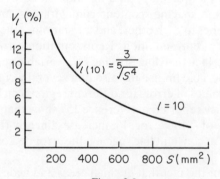

Figure 2.8

As a result, all or part of the lines will become somewhat shorter. If the error is with the negative sign, the length of the lines is increased by the value of ΔL_k. There is, however, no room for the lines to be lengthened, because when the contour is initially filled all the lines are drawn to reach the boundary. This problem has several solutions:

1. Increment ΔL_k is inserted in the space between the lines of the contour, if the gap between the lines, e, permits this.
2. Part of the lines filling up the contour are transferred into a senior (higher) weight p_{j+1}. The L_k' value is calculated from the formula

$$\Delta L_k' = \delta_\kappa \left(\frac{p_j^2}{p_{j+1}^2} + 1 \right) \tag{2.22}$$

3. Every line filling up the contour (for all the contours) is shortened by a certain percentage of length in accordance with coefficient c (e.g. when $c = 1.1$ the lines are shortened by 10 per cent). This makes it possible to free a space for the lengthening of lines. Then all the weights will increase according to coefficient c. The value of c is found from formula

$$c = \frac{\delta_k}{p_j \times \Delta L_k} \tag{2.23}$$

In this case, when calculating N_k in each contour it is necessary to take into account the alterations in weight values, i.e.

$$N_k = \sum_{i,k=1}^{n} L_{i,k} \times p_j \times c \tag{2.24}$$

After these corrections have been introduced, representation accuracy will depend on the total length of lines,

$$\sum_{i,k=1}^{n} L_{i,k}$$

and weight p_j, and for the last variant also on the number of lines n with the increment added, since such maps are composed automatically with the help of a computer and a graph plotter (see Chapter 5), the line plotting error will be determined by the value of discretization step, equal in present-day graph plotters to 0.01 mm. The total length of the lines filling up the contour will be equal to the sum of discretization steps of the graph plotter. Consequently, the rated error due to the rounding-off will only affect the last line. Its value will be equal to ± 0.005 mm. If the lines on the final compilation are drawn by a graph plotter with a cutter or a laser beam on a coated plastic, the anticipated accuracy of line engraving will be close to the rated accuracy of the computer. The printing defects of the map will be negligible since the smooth-delineation map is produced from the engraved final compilation by contact printing.

2.4.2. The method of representation by means of a continuous proportional cartogram

The method consists in applying equispaced lines whose thickness-to-interval ratio must be proportional to the density of discrete objects within a given territory. This ratio can be expressed by the following formula:

$$P = \frac{T}{l} c \qquad (2.25)$$

where P is the density of discrete objects, T is the line thickness, l is the width of the gap between the lines, and c is a coefficient [156].

To achieve the necessary representation accuracy in this method it is important to optimize the choice of the extreme line thickness and gap width values in accordance with the extreme values of the density of discrete objects. From formula (2.25) it follows that the greater the difference between the accepted maximum line thickness and minimum gap width values the more accurate the representation. Potential accuracy of the method is, however, limited by the dimensions of small contours.

The method gives an objective visual picture of quantitative relationships between the densities of discrete objects (Fig. 2.9). It should be noted that its principle has a certain similarity to that of the method of relief representation based on Leman's scale.

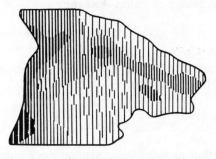

Figure 2.9

2.4.3. The method of representation by means of digital code signs

To represent discrete objects localized in points or on small areas other MRD modifications, characterizing more accurately their location in space, can be applied.

In reference 105 a method has been proposed to represent data in the form of digital code signs (DCS). The method permits showing in every point where the objects are localized or concentrated not only their actual (exact) quantity but also a number of additional characteristics, with the visual clarity of the occurrence of objects being preserved.

Figure 2.10

Figure 2.10 shows several variants of code signs. Variant (a) is a combination of three lines of equal length with a differing thickness, the thinnest line has the weight of 10, the medium one 100, and the thickened one 1000. Variant (b) has the same weight values, but to make the sign more compact the combination consists of half-length lines, i.e. it corresponds to half the weight value. In variant (c) the sign is presented in an even more compact form. It is possible to design code signs with lines subdivided into ten parts (variant d).

With the help of code signs one can show not only the total number of objects but also their qualitative characteristics. This is achieved by using different colours within the sign, or dividing it into several parts by small gaps in the lines [120].

Code signs can also be used to represent discrete objects localized on areas. In this case it is expedient to present DCS as segments of lines having a different length, distributing them within the area where the objects occur.

2.4.4. Combined use of new and traditional methods

1. Combining DCS with the traditional method of cartographic symbols makes it possible to represent data by the size of the symbol with preassigned intervals of discreteness. A stepwise scale of intervals is prescribed similarly to the traditional scale. The preferable form of a symbol is a square of a certain solid colour; for example only for urban settlements. The colour of the filling is used to denote qualitative parameters; for example, the ethnic composition of population, percentage of able-bodied population, or age composition (Fig. 2.11). DCS is superimposed on the background (filling) of a symbol. In this

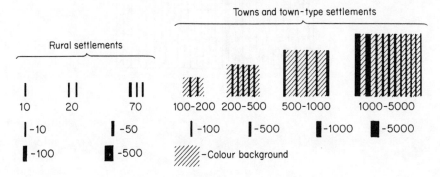

Figure 2.11

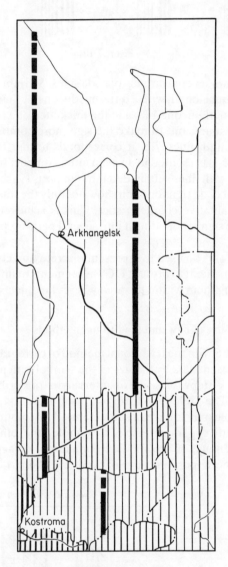

Figure 2.12

case, the length of line segments does not alter the weight. Weight is differentiated by line thickness. Line thicknesses are the same for rural and urban localities, but their weights differ, as expressed by different colour of lines.

2. Combining DCS with a cartogram can be used as follows:

(a) a traditional cartogram shows the average density of the distribution of discrete objects over an area, and DCS shows their quantity at certain points (local centres of concentration);

(b) a traditional or multiparametric continuous cartogram (MCC) shows the most scattered part of discrete objects, and DCS shows their quantity at certain points (e.g. rural population is represented by a cartogram, and urban population by DCS).

3. Combining MCC with a traditional cartogram is most expedient when distribution of discrete objects is represented within different boundaries. For instance, population density is represented by MCC within administrative boundaries, and by a traditional cartogram (colour background) within natural boundaries, or vice-versa.

4. Combining MCC with a diagrammatic map in the form of columns (Fig. 2.12) or with DCS (Fig. 2.13).

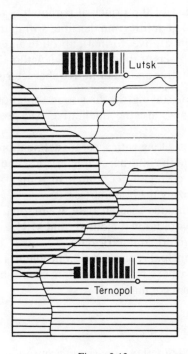

Figure 2.13

5. Combining the isoline method with MCC. MCC is given within the areas limited by the neighbouring isolines for a more accurate representation of discrete data.

6. Combining the isoline method with DCS. Other more complex combinations, consisting of more than two methods, can also be used on one map; for example, a combination of MCC and DCS with a traditional qualitative background or areal patterns. These can of course also represent other objects with non-discrete nature of distribution.

2.5. The Methods of Representing Continuous–Discrete Data

Continuous–discrete data, mainly describing areal objects (contours) and linear (extended) objects, depending on the purpose, scale, and content of the designed map, serve three basic representation purposes:

1. Geometrically precise reproduction of the boundaries of domains having one, two, or more interconnections, with the representation of their internal qualitative and quantitative characteristics.
2. Quasi-similar representation of the principal parameters of objects: area, shape, spatial localization, as well as internal qualitative and quantitative characteristics.
3. Integrated reproduction of the principal parameters of objects: absolute and relative total area, quantities and densities within the limits of assigned regions (natural, administrative, formal); absolute and relative total extension and sinuosity of objects or their boundaries within the limits of assigned regions; internal qualitative and quantitative characteristics.

The principal methods of solving the first representation problem, mainly related to large-scale maps, are the qualitative background and the areal patterns.

In both these techniques, along with accurate delineation of boundaries, it is also important to represent the internal characteristics, and there may be many of these, within one limited region (contour). In going to smaller scales, the problem arises of generalizing the boundaries.

Representation of continuous–discrete data, when solving the first problem, can be performed both by traditional methods and MRD. The latter has great potential in representing multiparametric data.

To solve the second principal representation problem, related to the reproduction of small objects, whose main parameters cannot be mapped to scale, the following methods are recommended:

(a) the multiscale method,
(b) the combined method.

The basic functions of these methods consist in accurate representation of the areas of contours or the areas and extensiveness of linear objects (e.g. rivers) with minimum possible losses of information on their localization and shape. They also make it possible to represent qualitative characteristics. These methods are mainly to be recommended for medium-scale maps, where large contours can be shown with a degree of generalization that preserves their natural boundaries, while the small contours, unrepresentable to scale, can be shown, with their main parameters being preserved, by means of quasi-similar transformations. A high detailedness and accuracy of data representation provided by these methods have become possible owing to MRD on which they are based.

When one is solving the third principal representation problem, mainly oriented at small map scales (when most of the objects cannot be mapped to scale) the following methods can be used:

(a) continuous multiparametric cartogram to convey integral characteristics of objects,
(b) pseudo-isolines (isopleths) for integrated representation of the basic parameters of contours or linear objects.

2.5.1. Representation of multiparametric data in the form of restricted connected domains

Geometrically precise and reliable representation of restricted connected domains (contours) is only possible on large-scale maps. In going to smaller-scale maps the question of contour boundary generalization becomes especially important.

Representation of multiparametric data in the form of restricted connected domains has to be based on the same principles as the traditional techniques of qualitative background and areal patterns. However, the problem of representing several characteristics (parameters) in the contour limits requires a non-trivial solution, which can be most optimally realized on normalized maps with the help of MRD.

Let us now consider different variants of using MRD to represent spatial data displayed within restricted connected domains as functions of several variables. Some representation variants (with qualitative characteristics only) coincide with the traditional definitions of the qualitative background and the areal pattern. Those containing quantitative parameters in addition to qualitative ones can be defined as the multiparametric qualitative background and the multiparametric areal pattern [120].

Connected areas, presented on a map in the form of contours filled with rasters, whose parameters correspond to different qualitative and quantitative characteristics of the represented objects or phenomena, can be expressed as the following function:

$$q_i = F(p_j(x, y)) \tag{2.26}$$

where $p_j(x, y)$, $(j = 1, 2, \ldots, m)$ stands for certain graphical parameters of rasters (line thickness, colour, etc.), and q_i ($i = 1, 2, \ldots, n$) stands for qualitative and quantitative characteristics of objects or phenomena corresponding to the raster parameters.

The simplest to represent are single-quality contours. These can be shown by rasters for the specific parameter with or without the outer contour line. Contours can have the form of simple or multiply connected areas.

The data are most commonly represented on a map in the form of contours with differing qualitative characteristics (e.g. a map of forest areas differing in the species composition of trees, or a map of agricultural lands differing in land-use). Qualitative characteristics of objects are best represented by the colour of lines or a combination of lines having a different colour within one symbol. Depending on the number of qualitative features, different combinations of raster line parameters are used.

Quantitative characteristics of objects can be conveyed by the thickness or the spacing of lines, as well as by combinations of line thicknesses. By combining line thicknesses one can construct digital signs; for example, to code the average height of trees on topographic maps.

With a large number of qualitatively different objects (several scores and more) being represented on a map, it is expedient to use combinations of colours and thicknesses in groups with free rearrangements of lines differing in colour and width.

There are instances in mapping when qualitatively different phenomena are interpenetrative, i.e. they intersect one another. Representation of such intersecting areas can be comparatively easy. Figure 2.14a shows a case of a single intersection of areas, $A \cap B = D_1$ and $B \cap C = D_2$, and Fig. 2.14b shows a case of multiple intersection, $A \cap B \cap C = D_3$.

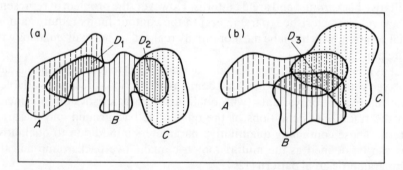

Figure 2.14

Forests constitute a typical example of natural objects represented in the form of intersecting areas. Characteristic of forests is the interpenetration of different tree species, i.e. the mixing of species with a simultaneous preservation of separate forest stands consisting of sufficiently homogeneous species. In this case the density of forest stand becomes a very important characteristic. It

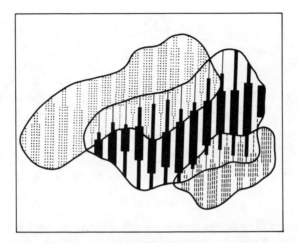

Figure 2.15

can be expressed by the thickness, and the tree species can be expressed by the colour, of lines (Fig. 2.15).

To represent the complex data of geological, geomorphological, pedological, and other maps use of combinations within groups in conjunction with a weak colour background (filling) is recommended. The solid colour background can represent the most general characteristics of a number of objects, i.e. their similarity and identity.

The presence of a solid colour background raises the information content and the readability of a map owing to the use of gaps between the raster lines. These should be printed with an optically dense (masking) paint so that their colour does not change under the effect of the background colour.

When designing raster signs it is necessary to observe the strictly logical principle of graphical expression: from the common to the individual. Let us consider this principle as exemplified by a geological map.

Each complex of uniform geochronological subdivisions will be regarded by us as a set of corresponding elements. We shall designate these elements as

M_1—group (era), G
M_2—system (period), C
M_3—series (epoch), O
M_4—stage, Y.

Each of M_i sets has m_j elements. The graph of quantitative relationships between the elements of all the sets for Cenozoic is shown in Fig. 2.16. It is seen from the graph that the following condition of the inclusion of sets is met:

$$M_4 \subset M_3 \subset M_2 \subset M_1$$

For a graphical realization of this hierarchical sequence of elements in a unified system of symbols we shall use the following notations.

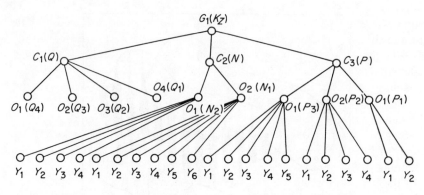

Figure 2.16

Groups (eras) will be designated by a weak solid colour background; systems (periods) by the colours of raster lines; series (epochs) by groups of lines, where one line stands for the first series of the corresponding system, two lines for the second series, three lines for the third, etc.; stages by the line thicknesses, i.e. the minimum width stands for the first stage of the corresponding series, the thicker one stands for the second, etc. To designate geological stages only the first line in a group will be used; i.e. only the first line in a group will have a variable thickness. So as to reduce the number of paints used in printing, geological systems can be designated by a combination of lines of different colour within a group.

2.5.2. The method of multiscale representation of areal objects (contours)

Raising the accuracy and detailedness of the representation of areal objects is of great importance, especially when they are small in size and widely scattered.

The most difficult is the representation of areal objects that cannot be mapped to scale. In traditional cartography such objects are either omitted or shown by an out-of-scale (point) symbol. Out-of-scale representation makes it impossible to estimate the actual area of objects and to get an idea of the contour shape. For widely scattered objects it is very important to represent their actual spatial occurrence, with their shapes and areas being preserved.

The method of raster discretization provides a positive solution to this problem, i.e. makes it possible to represent objects of practically any size (restrictions are imposed only in those cases when the density of objects is very high) without affecting their areal dimensions, the accuracy of localization and, to a certain extent, the shape of contours.

The technique proposed is based on the idea of developing a system of rasters having different weights with respect to area, where every line thickness (or colour) gradation step in the raster has, as it were, its own scale—larger than that of the map, except for the first step given to scale [120].

Let us assume that we have prescribed the following weights: $\bar{p}_1, \bar{p}_2, \ldots, \bar{p}_k$ with the raster line thicknesses of $T_1, T_2, \ldots, T_k$, respectively. For large contours that can be represented to scale, the weight of the first gradation step, distinguished by the largest line thickness (in our case, T_1), can be conventionally set equal to unity. Then the weight for contours that cannot be mapped to scale will decrease a certain number of times in accordance with the degree of the enlargement of contours, with the thickness of lines being correspondingly decreased.

When determining the weight values for gradation steps (expressed in units of area) one should take into account the spacing, thickness, and elementary length of lines taken as a discrete segment, e.g., 1 mm or the line advance (scanning step) of a scanner.

In the method discussed, when the data are being read by a machine, the elementary area is determined by scanner line advance d dividing the lines into discrete segments and, when the data are read visually, the discrete segment will be equal to the least division of a measuring device (e.g., a rule). The formula for the areal weight of first gradation step p_1 will therefore be expressed by the product of discrete line length d by the interval between raster axial lines and a square of scale denominator M^2:

$$p_1 = d \times l \times M_1^2 \tag{2.27}$$

The same can be written for other weight gradations, with the sole difference that they will have their own scale,

$$\left.\begin{aligned} p_2 &= d \times l \times M_2^2 \\ &\cdot \quad \cdot \quad \cdot \quad \cdot \quad \cdot \\ p_k &= d \times l \times M_k^2 \end{aligned}\right\} \tag{2.28}$$

The minimum value, l_{min}, determining the density of raster line depends on the sum of two quantities: the maximum line thickness, T_{max}, and the gap between the lines, e:

$$l_{min} = e + T_{max} \tag{2.29}$$

Taking into consideration the fact that the accuracy of data representation depends on the value of e, one should take it as small as possible, in accordance with the resolving power of human vision (if the map is designed to be read by man) and the resolving power of data medium (if the map is to be printed on paper it is necessary to make allowance for the whole publishing process).

This approach to determining the value of e is possible when a qualitatively homogeneous phenomenon is represented (in the form of qualitative background). When the problem is that of representing within one contour several phenomena by combining lines of different colour or by showing small inclusions of qualitatively different objects against the background of a qualitatively homogeneous large-area object, the value of l_{min} for the lines of the same colour (qualitatively homogeneous objects) is increased two, three, or more times, depending on the number of contours simultaneously intersecting as a result of their extension.

When the density of objects is not high, and they can be shown without intersections, the areal weights can be chosen arbitrarily, i.e. one may change not only the scale, M, but also the interval, l. Since, for larger weights, line thicknesses are taken smaller, it should also be expedient to take a smaller interval l, which would raise the representation accuracy.

This variant is especially appropriate when it is required to represent with sufficient accuracy the areas and shapes of small areal objects that cannot be mapped to scale. A characteristic feature of the method, from the point of view of the psychology of perception, is the estimation of an area not only by the size of a contour but also by its weight value, expressed by line thickness: the thicker the line (the more contrast and darker the contour), the larger the area; and the lighter the colour of a contour, the smaller its area.

Since in this method the area of contours, S, is determined by the total length of lines, obtained by summing up the lengths of individual vertical lines, L_q, i.e.

$$S = \sum_{q=1}^{n} L_q/d$$

with the areal weight accounted for, the procedure can, in a sense, be called the conversion of area into line length.

Such a method of solving the problem allows representing areal objects of, in principle, any size. This is achieved owing to the possibility of mutually inserting contour lines—one into another, i.e. into the spaces between the lines (Fig. 2.17b). Certain restrictions are, however, imposed on the possibility of contours insertion. In the case of a large density of objects a multiple intersection of contours occurs. This hampers map readability and makes it necessary to enlarge the weight values.

Measurements of areas on a map composed by means of this technique are easily performed with the help of dividers and a rule, as well as machines (automatic reading devices and computers).

Automatic computation of area S of a certain set of multiscale contours is essentially reduced to calculating the lengths of raster lines with their weights taken into account. If we designate a set of lines inscribed into these contours as Q, the formula from which the area is determined can be written as

$$S = \frac{1}{d} \sum_{q \in Q} L_q \times p_q \tag{2.30}$$

where L_q and P_q are the length and the weight of the qth line, respectively.

Specifically, Q can stand for:

(a) all the lines on the map;
(b) all the lines of a certain contour (then $S = (p/q) \Sigma_{q \in Q} L_q$);
(c) all the lines with a given weight.

Since the length of a line is determined as the product of scanner line advance d by the number of intersections N_i made on the raster line by the scanning element, the total area of contours with a different weight for any given region on the map can be represented by a simplified formula:

$$S = \sum_{q \in Q}^{n} p_q N_q \qquad (2.31)$$

The accuracy of measuring the areas of contours having different weights, whichever technical means of measurement are used, will be markedly different, this difference being directly dependent on the areal weight taken for a specific contour. Quite obviously the highest accuracy of measurements will be attained for the contours with the smallest areal weight, and the contours with larger weights (i.e. those represented to scale) will be measured less accurately. Because of this their relative errors will be close in magnitude. To increase the accuracy of representation and the visual clarity of images for areal objects represented to scale it is advisable to indicate the boundary lines of contours (see Fig. 2.17b) with a solid colour instead of raster lines inside, and to measure the area by the contour boundary.

2.5.3. The multiscale method of representing rounded objects

Among rounded objects one can mention craters, sink holes, man-made pits, etc. Widely occurring rounded objects are the craters of celestial bodies.

Craters are among the basic substantive elements of the topographic maps of the near-Earth celestial bodies. It is therefore necessary to attach great importance to the accuracy and detailedness of their representation. The multiscale method satisfies these requirements in the best possible way [120].

The shape of rounded objects makes it possible to show every object with the help of only one line whose length corresponds to its diameter (Fig. 2.18a).

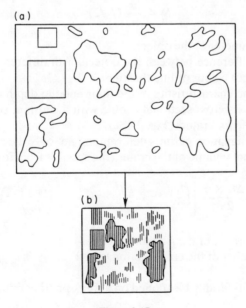

Figure 2.17

72

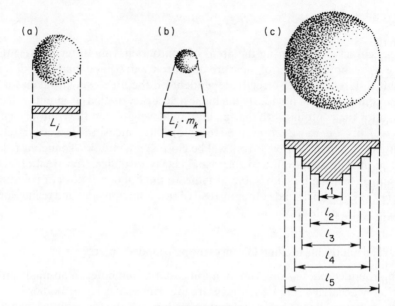

Figure 2.18

The objects that cannot be mapped to scale are represented by means of various scale factors, $m_1, m_2, \ldots, m_k$, increasing the length of their diameter (Fig. 2.18b). The area of an object in this case can be written as

$$S_i = \frac{\pi}{m_k^2} (L_i/2)^2 \qquad (2.32)$$

where L_i is the diameter of the object.

The outward difference between the objects with different weights can be shown by a different colour or line thickness.

To represent the characteristics of the floor and the depth of an object it is expedient to give its cross-sectional profile with a constant step Δh, which is especially desirable for craters (Fig. 2.18c).

This kind of designation permits determining not only the area of an object but also the internal volume of the crater (the spoil) V from formula

$$V = \sum_{j=1}^{n} \frac{\Delta h \times \pi}{2m_k^2} \left[\left(\frac{l_1}{2}\right)^2 + 2\left(\frac{l_2}{2}\right)^2 + \cdots + 2\left(\frac{l_{n-1}}{2}\right)^2 + \left(\frac{l_n}{2}\right)^2 \right] \qquad (2.33)$$

where l_j is the diameter of the jth cross-section,
j is the number of the cross-section ($j = 1, 2, \ldots, n$).

The lines should preferably be applied on a weak photographic background or relief shading.

2.5.4. Combined methods of representing areal objects

When the concentration of areal objects is very high, which is characteristic for the compilation of small-scale maps, it becomes difficult to implement the multiscale representation method, because of the multiple mutual intersection of extended contours, i.e. the contours that have been transformed into scales larger than that of the designed map. For this case the following technique of representing small areal objects is proposed.

The group of the initial map contours whose concentration is inadmissible for multiscale representation are circumscribed by one common contour, and the other dispersed contours are represented with the help of the multiscale method (Fig. 2.19). The boundary of the joint contour formed in this way is plotted on the composed map. The data on all the small contours falling within the joint contour are represented by a system of MRD raster lines characterizing certain assigned parameters of objects. The choice of parameters is determined by the purpose of the map. The system of raster lines can accordingly be calculated and constructed in different ways. Let us now examine some of them.

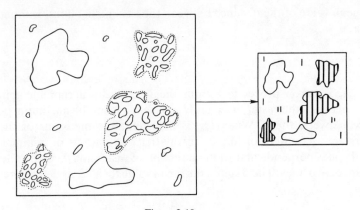

Figure 2.19

2.5.4.1. The mode of representing the absolute value of the total area of small contours, and its percentage relative to the area of the joint contour

In this mode it is imperative to measure the area of the joint contour and the total area of its constituent small contours. The problem is then reduced to calculating the spacing of lines, i.e. determining the distance, l, between the axial lines of the raster (in accordance with the accepted weight, p). Proceeding from the idea lying at the basis of the method of multiparametric continuous cartogram, parameter l can be calculated from the formula:

$$l = \frac{p_k}{\delta_i \times M^2 \times 0.01} \qquad (2.34)$$

where p_k is a weight expressed as the number of units of area per 1 mm of line length; k is the weight number; M is the denominator of the scale of the composed map, in km; δ_i is the percentage of the area occupied by small contours relative to the area of the joint contour, S_i.

This value is found from the formula:

$$\delta_i = \frac{\sum_{j=1}^{n} S_j}{S_i} \times 100 \tag{2.35}$$

where S_i is the area of the jth small contour within the joint contour $(j = 1, 2, \ldots, m)$.

Substituting the δ_i value into formula (2.34), we shall have

$$l_i = \frac{S_i \times p_k}{\sum_{j=1}^{m} S_j \times M^2} \tag{2.36}$$

When only one weight is taken, which is the most probable case for this mode, its value can be found from formula

$$p = \bar{l} \times \bar{\delta} \times M^2 \times 0.01 \tag{2.37}$$

$\bar{l}$, the mean value of the distance between the raster axial lines, is found from formula:

$$\bar{l} = \frac{l_{max} + l_{min}}{2} \tag{2.38}$$

where l_{min} is taken proceeding from the graphically acceptable minimum distance between the raster lines, and l_{max} is taken from the maximum possible distance between them. When l_{max} is determined, dimensions of the joint contours are accounted for in such a way that the contour of minimal width (along the axis perpendicular to the direction of raster lines) can be filled by a minimum of two lines. The $\bar{\delta}$ value, the mean value for all δ_i, is determined from formula:

$$\bar{\delta} = \frac{1}{n} \sum_{i=1}^{n} \delta_i \tag{2.39}$$

where n is the number of joint contours.

In the proposed mode of representing contour objects, one weight can be recommended since, owing to the non-combined small contours being shown by the multiscale method, the line spacing variations, l_i, will fall within the assigned limits, l_{min} and l_{max}. The weight value is rounded off to a suitable quantity (the rounding-off does not affect representation accuracy). For the convenience of determining the exact values of the percentage of total area occupied by small contours relative to the area of the joint (generalized) contour with the help of the simplest means we recommend constructing a nomograph. The percentage is plotted at an arbitrary scale on the horizontal axis, and divisions in millimetres on the vertical axis. When plotting the curve the percentages, obtained proceeding from δ_{min} and δ_{max}, are taken from the horizontal axis and substituted into formula (2.34), after which the corresponding l values are calculated.

The total value of the area,

$$\sum_{j=1}^{m} S_j$$

of the small contours, lying within the joint contour is found as the product of the weight by the measured total length of all lines filling these contours:

$$\sum_{j=1}^{m} S_j = p \times \sum_{q=1}^{n} L_q \qquad (2.40)$$

where L_q is the length of the qth line, mm;
$\quad n$ is the number of lines, ($q = 1, 2, 3, \ldots, n$).

To facilitate prompt visual estimation of the percentages of joint contours a scale similar to the one shown in Fig. 2.6c is plotted.

Large contours that can be plotted to scale are shown within their actual boundaries. Small curvatures in the boundaries of contours are generalized in accordance with preassigned criteria so as to preserve the equality of the contour area, since this is the basic parameter in this mode.

2.5.4.2. Representation of the total number of contours falling within the joint contour and of their statistical density (the ratio of the number of contours to the area of the joint contour)

In this mode the basic parameter is the nature of the spatial distribution of contours, i.e. what is emphasized here is the statistics of contours distribution rather than their area. All calculations and plotting are performed based on the same principle as in the method of multiparametric continuous cartogram. In this case, symbol N in formula (2.8) denotes the number of contours, and symbol S denotes the area of the joint contour. Contours representable to scale are shown as in the previous mode, and those that cannot be represented to scale are shown by out-of-scale short segments of lines, with one line segment assigned to one contour.

2.5.4.3. The mode of representing the statistical density, quantity and percentage of the total area for small contours

Similarly to the previous mode, a system of lines is designed and plotted to represent the statistical density and the number of small contours within the joint contour. The third parameter—percentage of the total area occupied by small contours—is represented by the interval between the lines. The interval, l_i, between the lines is taken for 100 per cent, irrespective of their density (i.e. the distance between lines, l).

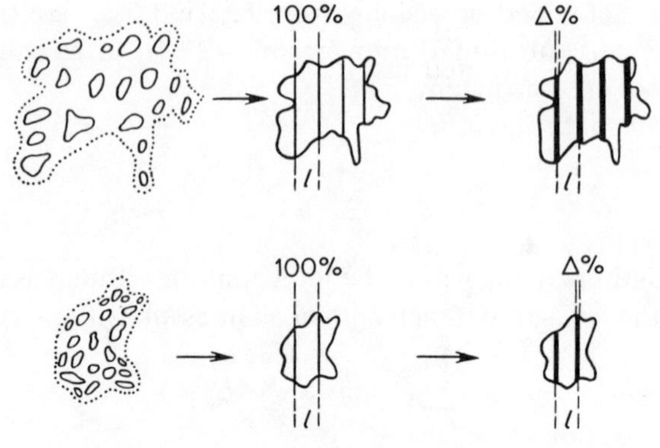

Figure 2.20

Percentage $\delta_i\%$ of the total area of the contours falling within a certain joint contour is found from formula (2.35), and then line thickness T_0 is determined from the formula: $T_i = l_i(\delta_i/100)$. One can calculate the value of T_i and then plot the lines with the help of an interpolating transparency or a computer (Fig. 2.20). Thus, by the ratio of the line thickness to the blank space one can judge the percentage of the total area of small contours relative to the area of the joint contour, and by the density of these bands one can judge the statistical density of contours and their quantity. The latter can be obtained with the help of a special nomograph, as described for the first mode.

For the percentage of the total area of contours to be represented with sufficient accuracy and clarity it is very important to find the optimum values of l_{min} and l_{max} when determining the weight value, p_k.

Small scattered contours that cannot be mapped to scale and lie outside the joint contour can be shown with the help of the multiscale method, or based on the principle of raster line segments having different weights, which will be discussed below.

Small scattered contours that cannot be mapped to scale and lie outside the joint contour can be shown with the help of the multiscale method, or based on the principle of raster line segments having different weights, which will be discussed below.

All the small scattered (uncombined) contours unrepresentable on the map are subdivided into n areal gradations. The difference between these gradations is shown by line thickness at a constant length of line segments, the line thickness increasing with area. For instance, at a scale of 1 : 1,000,000 the contours with an area of up to 5 ha are shown by line segments whose thickness is 0.1 mm; from 5 to 10 ha, 0.25 mm; from 10 to 20 ha, 0.5 mm; from 20 to 50 ha, 0.8 mm; and the contours with an area larger than 50 ha are represented to scale (Fig. 2.21). If it becomes necessary to show the contour areas more precisely one can introduce a scale along the line length. A limiting length of the segment

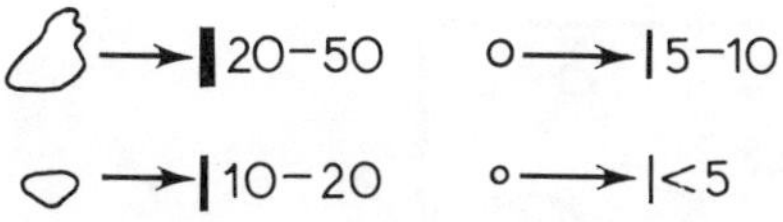

Figure 2.21

(e.g. 5 mm) is assigned to be taken equal to 100 per cent for each gradation. Areal data are represented with a certain preassigned accuracy determined by the degree of discreteness, e.g., equal to 20 per cent (accuracy ±10 per cent). Then a contour with an area equal, e.g. to 7 ha will be represented by a segment 2 mm long and 0.25 mm thick, a contour with the area of 16 ha by a segment 3 mm long and 0.5 mm thick.

The colour of line segments, representing the scattered contours, can be the same as that of the lines filling the joint contours.

2.5.4.4. *The mode based on summarized representation of the parameters of the modes described above*

Mutually perpendicular arrangement of lines makes it possible to show all the characteristics represented in the joint contours by means of the above modes. In other words, indices of one mode are shown by vertical lines, and of the other by perpendicularly lying horizontal lines. Though the number of indices is sharply increased in this case, the graphic properties of the map are preserved. Moreover, this kind of representation gives a clearer, more objective picture of the spatial distribution of contours, as it conveys an integrated characteristic of the density, the area, and the quantity of contours. True, selective readout of data on a specific parameter is more difficult than with a map containing indices represented by means of only one of the above modes. However, a map combining two such modes carries an incomparably greater amount of information.

All the above modes are designed for multilevel scale generalization, where the accuracy of contours representation is diversified. With such generalization the most accurately and reliably defined are small scattered contours represented by the multiscale method. Precise representation of small scattered contours is of great significance in some cases (e.g. in investigating the spatial distribution of ore bodies in geology). But for certain purposes, requiring mutual correlation and visual comparison of small contours, this kind of representation may not satisfy the map-user. In this connection we shall examine another technique, where all the small contours unrepresentable to scale are shown in the same way (multiscale representation of small contours is excluded). All the mapped territory is subdivided, by the degree of the concentration of contours, into a number of regions presented in the form of joint contours, except for the contours that can be represented to scale

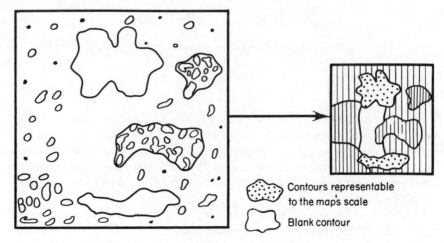

Figure 2.22

(Fig. 2.22). These regions can be differentiated by different features (characteristics). If the statistical density of contours is shown, then the parameter forming the basis on which the regions are singled out is the average distance between the weighted centres of contours or the ratio of the number of contours to the region area. If areal density is shown, then the relevant parameter is the degree of areal coverage of the region by the contours, i.e. the ratio of the area occupied by contours to the free area or the total area of the region. The procedure of indicating the regions and determining their values (assigned ratios) can be performed visually without any special calculations and measurements (if the map is composed manually), and, when a computer is used, by calculations for square or hexagon-shaped cells (Fig. 2.23a). Within the limits of each cell a value is found for one of the assigned ratios, e.g. for areal density, from formula (2.36). Then the proximity criterion for the values obtained in the cells is established in accordance with a certain scale of intervals. This measure is used to combine cells into regions; that is the cells whose values fall within one interval are combined and designated in accordance with the gradations (intervals) of the scale (Fig. 2.23b). The boundary of a region is plotted following the boundaries of joint cells in the form of a smoothed-out line (see Fig. 2.23c). Within the limits of every region obtained in this way the mean value of $\bar{\delta}$ is found, and then using formula (2.36), if the map of the areal density of contours is compiled, or formula (2.8), if the map of the statistical density of contours is compiled, the distance, l, between lines is found. After this, the regions within their boundaries are filled with lines corresponding to the computed values of l_i. The exact localization of small contours can be shown by dots (filled-in circles) in their weighted or geometrical centres.

If small contours on the original map are qualitatively different (e.g. for different species of trees) and their qualitative characteristics have to be shown not in the aggregate but individually, the following technique can be applied.

Within each region are singled out (when this is feasible) subregions where

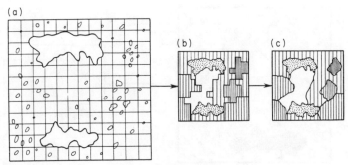

Figure 2.23

one of the qualitative characteristics is predominant. Within the subregion the prevailing parameter is shown by one or two colours and the other characteristics are given in the aggregate. Their simultaneous representation can be achieved in the following ways.

A. One line (regardless of its length) is conventionally taken for a certain percentage, e.g. 10 per cent. Then a group of neighbouring lines, in our case ten lines, will account for 100 per cent. In accordance with the calculated $\delta\%$ values the corresponding number of lines representing the predominant characteristic is filled with a chosen colour and the others are either not filled at all or have a different colour (Fig. 2.24a).

B. When mode A is inappropriate (a region has to contain no less than ten lines), every line, irrespective of its length, is taken for 100 per cent and divided into proportional segments of a certain colour (Fig. 2.24b).

C. Out of all the l_i values, obtained after being calculated for all the regions, the minimum value, l_{min}, is identified and taken for 100 per cent of line thickness dependent on the $\delta_{max}\%$ index. When the predominant qualitative characteristic is represented, each subregion will have its own line thickness, $t_i = l_{min} \times (\delta_i/100)$, calculated from $\delta\%$. The whole subregion is then filled with lines of T_i thickness with the distance between axial lines equal to l_i (Fig. 2.24c).

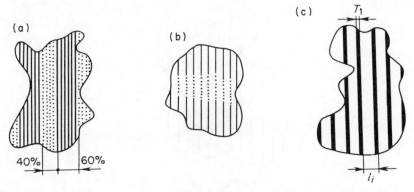

Figure 2.24

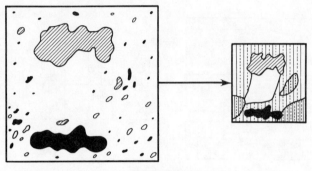

Figure 2.25

In principle there is a possibility of avoiding the use of subregions by showing the percentages of several qualitative characteristics within a region. To do this one can apply the same modes (A, B, or C), with the only difference that it will be necessary to use a greater variety of colour hues, i.e. assign its own colour for each qualitative characteristic shown (Fig. 2.25). The principles applied to diagrammatic maps can also be used. All the lines within a region can have the colour of the prevailing qualitative characteristic, with the exception of its middle part where some of the lines, in accordance with the weight taken for a given region, are presented in the form of a rectangle (Fig. 2.26). It is necessary that within the rectangle thus formed expression

$$p \sum_{q=1}^{n} L_q$$

should be equal to the sum of all the other qualitative characteristics. Relationships for the total values of areas with respect to each characteristic can be given inside the rectangle in percentages and absolute quantities. Modes other than those described above (A, B, and C) can also be used.

In all the above-mentioned techniques the emphasis was placed on representing the contours that cannot be mapped to scale. In the practice of small-scale

Figure 2.26

mapping of areal objects one may have to compile maps on which it is impossible to show contours (even the largest ones) to scale.

In this case it is expedient to use the cartogram technique. When the requirements for representation accuracy are high the method of multi-parametric continuous cartogram is recommended. Areal density can be represented with the help of the first mode, the only difference being that, instead of the boundaries and areas of joint contours, the boundaries and areas of administrative territories (e.g. economic regions, districts, etc.) or of specifically chosen natural or economic territories are taken. Qualitative characteristics can be represented similarly to modes A, B, and C in conjunction with the cartogram technique. In this case we are obviously dealing with special-purpose generalization, since a quite definite end is pursued here: to show in the most generalized form the distribution of objects with respect to the assigned parameter. Some of the above techniques (with the exception of the multiscale method) constitute a combination of special-purpose and scaled generalization, as they include identifying the boundaries of the areal or statistical distribution of small areal objects by the degree of their concentration or, in other words, their density.

2.5.5. The multiscale method of representing linear (extended) objects

The width of a large number of linear objects, for example small and medium-sized rivers, ravines, geological faults, or rifts, cannot be mapped to scale, and in traditional cartography they are usually shown by conventional symbols. Out-of-scale representation of linear objects precludes quantitative estimation of the data on the width of a river and its area. This seriously restricts the effectiveness of using maps to solve the problems of engineering geology, geological prospecting, etc.

Let us assume that the whole channel of a river shown on the original large-scale map is subdivided crosswise into discrete sites following one another at a constant spacing d along the water course. The smallness of discretization step allows us to write that the area limited by the width of the river image, C_i (plotted to scale), at the ith point with discretization step d is $S_i = d \times C_i$. Then the actual area of the whole assigned jth section of the river is the product of the sum of these areas (sites) by the square of scale denominator.

$$S_j = \sum_{i=1}^{n} d \times C_{ij} \times M_0^2 = \sum_{i=1}^{n} S_{ij} \times M_0^2 \qquad (2.41)$$

To represent these parameters for the rivers that cannot be mapped to scale we shall introduce different scale factors. Factor k_1, for rivers representable to scale, we shall equate to unity. All the rivers will be broken up into several sections (gradations) of decreasing width, and each gradation will be assigned its own scale factor $k_1, k_2, \ldots, k_m$. As a result we shall have a scale containing m gradations:

$$\left.\begin{array}{cccc} C_{11} & C_{21} & \cdots & C_{n_1 1} \\ C_{12} & C_{22} & \cdots & C_{n_2 2} \\ \cdots & \cdots & \cdots & \cdots \\ C_{1m} & C_{2m} & \cdots & C_{n_m m} \end{array}\right\} \qquad (2.42)$$

where n_i is the number of sites in the ith gradation.

Using this scale we can perform the transformation of rivers by their width into what is actually a multiscale representation, since the product of the variable k_j factor by the represented width of the river, C_{ij}, changes the scale of river representation on the map in accordance with the scale gradations taken. Rivers or their sections belonging to a certain gradation will be designated by colours (paints) of a differing degree of brightness (luminosity), the larger the scale factors (i.e. the smaller the river) the brighter the colour. Discernibility of coloration brightness must be sufficient to allow reliable identification.

When composing a map it is unnecessary to calculate the width of the river for each discrete site. It is sufficient to determine it at the points of channel width variations and then connect the calculated points by generalized lines along the river banks (Fig. 2.27).

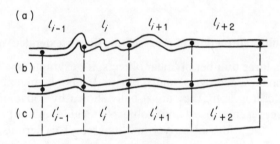

Figure 2.27

At a constant scale along the river channel the area of any single section (site) of the river, or a group of such sections, belonging to the jth gradation with scale factor k_j, can be determined from formula

$$S_j = \frac{1}{k_j} \sum_{i=1}^{n_i} C'_{ij} \times d \times M_1^2 \qquad (2.43)$$

where $C'_{ij} = C_{ij} \times k_{ij}$.

The total area of a river, or a group of rivers, belonging to different gradations can be calculated from the following formula:

$$S = M_1^2 \times d \times \sum_{j=1}^{m} \left(\frac{1}{k_j} \sum_{i=1}^{n_i} C'_{ij} \right) \qquad (2.44)$$

In the course of map compilation when one passes to smaller-scale maps, generalizations of meanders are inevitable. This is known to result in reducing the river channel length, the length of the river at different sites having different errors because of the non-uniformity of meandering. At the same time this will also inevitably affect the accuracy of determining the total area of a river. So as to preserve this parameter and account for it in determining not only the river length but also its area, the following procedure is recommended [133].

On the large-scale original the river length is measured, with the points corresponding to the conventionally taken at the map scale discrete distance l^0 (e.g. 1 km) being fixed along the river channel. When generalizing the meanders these points are transferred (projected) onto the new river channel of the composed map in perpendicular direction. As a result, on the composed map these points along the river channel will be located at a different distance l' from one another (Fig. 2.27). Reduction of the distance between points, relative to initial distance l^0, will be indicative of the presence of generalized meanders. The condition of $l^0 \times M_1 \geqslant l' \times M$ has to be observed in this case. The smaller the distance between points the greater obviously is the number of generalized meanders. This technique makes it possible to determine with sufficient accuracy the length of the river channel and to introduce corrections into area determinations. The location of the terminal points of discrete segments l'_i can be shown on rivers represented by two lines in the form of dots, and on those represented by one line they can be shown as gaps in the river channel.

The length of any section of the river, ΔL, and of the whole river course, L, will be determined from the following respective formulae:

$$\Delta L = M l^0 \left(\frac{\Delta L'_1}{l'_1} + n + \frac{\Delta l'_2}{l'_2} \right); \quad L = M \times l^0 \times N \tag{2.45}$$

where n is the number of discrete segments lying between the terminal points of the line length measured along the river channel, $\Delta l'_1$, $\Delta l'_2$ are the parts of the discrete segments falling respectively at the beginning and the end of the assigned segment ΔL (Fig. 2.28), N is the number of the discrete segments along the whole river course.

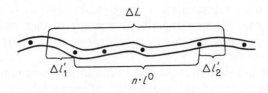

Figure 2.28

The area of the river section accounted for by one scale factor can be determined from the formula:

$$S = M^2 l^0 \frac{1}{k} \left(\frac{\Delta l'_1}{l'_1} C_1 + \sum_{i=1}^{n} C_i + \frac{\Delta l'_2}{l'_2} C_2 \right) \tag{2.46}$$

The total area of the whole river channel with all scale factors accounted for can be expressed by the following formula:

$$S = M^2 l^0 \sum_{j=1}^{m} \frac{1}{k_j} \sum_{i=1}^{n_j} C'_{ij} + \frac{\Delta l'_2}{k_j \times l'_2} C_2 \qquad (2.47)$$

where j is the number of scale factor ($j = 1, 2, \ldots, m$).

The initial stage in constructing a multiscale image is the classification of rivers by their width. The classification consists in placing rivers or channel sections into groups in accordance with the preassigned intervals of river widths conforming to the accepted scale of gradations.

To simplify the selection process it is recommended that the mean width of the river channel be determined from several values running in succession:

$$\overline{C}_p = \frac{1}{n'_p} \sum_{i=1}^{n'_p} C_i \qquad (2.48)$$

where p is the number of a section ($p = 1, 2, \ldots, N$). In other words, river courses are divided into sections having n'_p points each, then the mean width values are determined from the values obtained in the preceding operation, and these are grouped into gradations with the fixation of the ends of the sections. The length of a section, or, in other words, the number of channel width values running in succession, from which $\overline{C}_p$ is calculated, is determined on the basis of the scale of the compiled map, its purpose, and taking into account the region where the mapping has been performed.

2.6. The Methods of Representing Continuous Data

Continuous data in the form of a function of two variables, $z = f(x, y)$, are represented by three basic modes:

1. continuous–discrete representation using the method of the section of function relative to the variable z at a constant (with a few exceptions) Δz increment;
2. discrete–continuous representation using the method of subdividing the $z = f(x, y)$ function into n functions in which certain fixed values of planimetric coordinates x or y (e.g. $z_{x_1}(y), z_{x_2}(y), \ldots, z_{x_n}(y)$) are taken as the variable quantity;
3. continuous–continuous representation of the surface by a family of curves every one of which corresponds to a function of one variable, $z_{x_j}(y)$ or $z_{y_i}(x)$. Complete continuity is achieved at an infinitesimal discretization step.

In the first representation mode the following techniques are used:

(a) isolines, when the function is assigned by a family of curves of the same level, $z_i = f(x, y) = $ const., with a constant Δz increment (the standard isolinear representation method);

(b) layered isolation (by colour or texture) of intermediate zones limited by $z_i = f(x, y)$ lines of equal level with a constant or variable Δz increment (the standard hypsometric representation method).

The second and the third representation modes, according to the generally accepted terminology, are placed in the category of relief diagrams (block diagrams). The second mode is characterized by a unidirectional series of curves (profiles) or, in other words, traces left after the section of the surface or the complex structure of an object (model) by vertical parallel planes, for example the layered structure of the lithosphere represented in the form of a block diagram in the axonometric, perspective, or some other projection. The third mode serves the same representation purposes, the difference being that the series of profiles (family of curves) or parallel vertical planes of the section are mutually perpendicular, i.e. one series corresponds to the function of $z_{x_j}(y)$ variable and the other corresponds to $z_{y_i}(x)$.

Block diagrams are most widely used in geology to show the structure of the Earth's surface in a vertical section. They are also used to represent the interrelation between the relief and the soils, the geophysical structure of the entrails of the Earth, the occurrence of water masses, deep-sea currents, etc.

A block diagram is a peculiar form of cartographic data representation. As a rule it is constructed based on a map. It is thus the result of converting an isolinear image on a map into another spatial representation of continuous data (inverse conversion is also possible).

Continuous data can also be represented with the help of MRD. Its use is particularly effective when it is necessary to simultaneously represent a number of parameters, i.e. to show on one map several qualitative and quantitative characteristics in combination. In such cases the best results are obtained by combining MRD techniques with isolines.

2.6.1. The isolinear–numerical method of representing continuous surfaces

For the representation of continuous data (e.g. a gravitational field, magnetic field, relief, etc.) on normalized maps of enhanced accuracy and detailedness (information content) the *isolinear–numerical method* is recommended [133]. It is based on combining the isolines, normalized by means of a latent optical code, with digital code signs (DCS).

Digital code signs can represent initial (primary) numerical data, for example the data on the geodetic control points obtained during relief surveys. Invisible DCS images are applied in the gaps between isolines (with luminescent white paint). They are given in the form of reduced numerical values, as an increment to the isoline of lesser value, $\Delta H' = H' - H_i$, where H' is the value at a point, H_i is the value of the lesser isoline. Variation of $\Delta H'$ values lies within the section step, H, i.e. the condition of $H_i + \Delta H' \leq H_{i+1}$ is met. Owing to this the DCS is compact, consisting of a small number of code elements and thereby

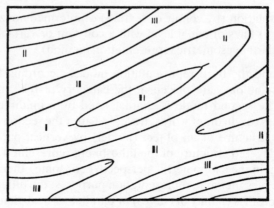

Figure 2.29

occupying a small area on a map (Fig. 2.29). DCS can be constructed in different ways, the simplest being a system of Roman numerals (Fig. 2.30).

Serving as a graphical information carrier, such a normalized map can be read by a machine with the help of optical electronic blocks of the scanning device and a computer, by decoding the latent optical codes and identifying their values in a unidimensional space from a sequence of optical signals produced by the isolines with different spectral characteristics [111]. Being a digital data medium, the normalized map is mainly designed to be read by a computer. However, if necessary the digital data can also be read visually by man. In both these cases the map is exposed to an ultraviolet radiation source. The latter, by stimulating the luminescence of the digital code signs, makes

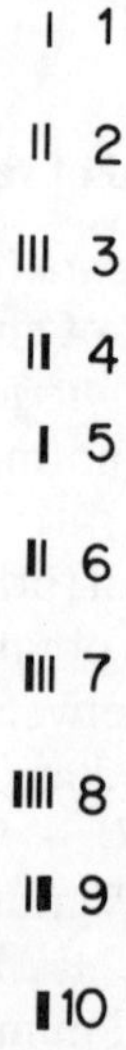

I 1

II 2

III 3

II 4

I 5

II 6

III 7

IIII 8

II 9

I 10

Figure 2.30

them visible to man and perceptible to the photoelectronic sensor elements of the scanner. The numerical value of a point is determined as the sum of the lesser isoline elevation value and the digital code value, i.e. $H' = H_i + \Delta H$.

This kind of representation has a significant advantage over traditional representation: preserving the visual clarity and the shape of surface elements it can convey information without losses and with an accuracy (on large-scale maps) close to that of the initial (primary) data with a highly reliable identification of data values by a machine operating in a fully automatic mode.

CHAPTER 3
Latent Coding of Images on Maps

The merits of some traditional data representation methods, from the point of view of their metric qualities and representation accuracy (e.g. the method of isolines), make it necessary to find ways of solving the problem of automatic map readout that would basically preserve the traditional form of cartographic image and, at the same time, make possible a simple procedure for reading and processing it by machine.

The most expedient way of satisfying these requirements consists in constructing images with a latent code, designed for machine reading. In this case, as we shall see later, latent coding of cartographic images on a map opens up new possibilities in terms of data representation.

This chapter describes the newly proposed methods of latent coding which make it possible to simplify radically the procedure of code reading by a machine, with the traditional outward form of data representation being preserved.

The main idea of the latent coding of images on maps lies in constructing the image in such a way that the map preserves its customary outward appearance and, at the same time, carries some latent data intended to be read by a machine.

The two most promising methods of the latent coding of images are:

1. optical coding based on the use of luminophors;
2. magnetic coding based on the use of ferromagnetic materials.

The idea of a latent code based on the use of luminophors has already been realized and applied in actual practice.

This chapter deals exclusively with the problems of using luminescent properties for the optical coding of images as applied to computerized and visual reading of cartographic data. The methods of optical coding described in this chapter have been designed for the most complex, from the point of view of automated readout, data; viz. three-dimensional data represented in the form of isolines and stereoscopic models, as well as two-dimensional linear data (extended objects) represented in the form of various curvilinear images. Such images are constructed with the help of special paints with luminescent fillers and, in some cases, special kinds of paper.

The second method, based on the use of ferromagnetic materials, is not considered in this chapter since no experiments with it have been performed for cartographic purposes, although it has gained practical application in other fields of engineering cybernetics.

A magnetic head is used to read the images applied with ferromagnetic paint. It is obvious that this method can be used to construct coded images on maps.

3.1. Analysis of Luminescent Properties as Applied to Latent Coding of Data on a Map

One of the most complicated questions in the problem of automating the readout of data from maps is known to be that of identifying objects on maps, especially by their qualitative characteristics. The use of luminescence on maps simplifies this problem considerably. Luminescence is used to construct special luminous images on a map, which simplify the logical procedures and the technological implementation of selective readout of cartographic data performed by a machine. The process of producing luminescent maps is not confined to simply replacing (in the printing of maps) an offset paint with a luminous one, as done, for example, to produce air navigation maps (where it is necessary to read a map in darkened conditions with an invisible source of ultraviolet radiation), but is aimed at plotting a special coded luminescing image on the map.

Let us consider the principal properties of luminescence that are of particular interest from the point of view of the automatic reading of cartographic information.

Owing to the fact that the phenomenon of luminescence is not manifested by all substances there is a possibility, when a map is being read, of reliably distinguishing the special information carried by the map from its auxiliary elements. This means that substantive elements of a map which are not supposed to be read by a machine, e.g. geographical data, mathematical base, can be printed with a non-luminescing paint, while a luminous paint is used to print the special elements of the map meant for machine reading. Spatial fixation of the automatically read data can either be achieved visually or coordinated automatically at the moment of reading. It is desirable (in some readout methods) to eliminate the luminescence of paper, observed as a rule in the short-wave radiation region, or to make use of special non-luminescing paper. On a map produced by this technology it is not difficult to identify, with an automatic device, the luminescing image (under ultraviolet radiation) and to make the necessary measurements, separately from the auxiliary (non-luminescing) elements which are not designed to be read and processed automatically. All the elements can be read out visually in their interrelation (if the data designed for automatic readout are not latent) in natural lighting. This property of luminescence should be regarded as one of the most important as applied to maps.

Another very important property of luminescence, in its application to maps, is characterized by the Stokes–Lommel law. According to this law the emission spectrum as a whole, and its maximum, are shifted towards the long-wave region relative to the absorption spectrum. Because of this property of luminescence, selective readout of cartographic data from the images of qualitatively different phenomena can be performed by means of differentiated excitation, i.e. by using radiation sources having shorter wavelengths than the fluorescence of measured objects, with the luminescence being recorded by transducers equipped with filters that do not retain luminescence with a maximum at longer wavelengths. The property of luminescence to transform light, i.e. convert radiation of one wavelength into radiation of another wavelength, is extremely valuable in the automatic data readout from maps.

The next important property of luminescence is governed by the law of the independence of emission spectrum from the wavelength of excitation light. This considerably simplifies the measurements on a luminescent map, because broad spectral regions can be utilized.

An important property of luminescence, of particular interest in solving the problem in question, is the duration of persisting luminescence. The luminescence persistence of different substances varies within very wide limits.

Substances are known with persistence amounting to only several billionths of a second, and there are some whose persistence reaches several hours and even days. By using paints (inks) having different luminescence persistence durations we can, with the help of special devices, raise the number of reliably identified symbols on a map. Medium and prolonged persistence are basically characteristic of crystallophosphors. No offset printing methods with the use of crystallophosphors have as yet been developed. The use of luminescence of differing persistence for the reading of cartographic data will find a limited application, i.e. at present it is only feasible in cases where the image is applied manually or with stencils.

Luminescing maps are mainly produced with the help of luminescent paints in a mixture with organic luminophors (luminors).

3.1.1. Comparative analysis of the possibilities of computer selection for images printed with conventional paints and those having a luminescing filler

To make the expediency of using luminescence more obvious let us compare the possibilities of automatic separation of qualitatively different data on a luminescing and a conventional map. There exists a major difference between luminescence and reflected and diffused light. Irrespective of the direction of excitation light, the resulting luminescent light propagates uniformly in all directions, which is a substantial positive factor in machine-assisted readout of data from maps. In distinction to luminescence, the light reflected from the map surface propagates non-uniformly because of the incomplete diffuse reflection by paper (diffuse reflection accounts for about 60–70 per cent of the light).

The luminous flux reflected from every painted surface upon illumination by a source with an equienergy spectrum constitutes an essentially continuous spectrum, including the infrared and ultraviolet regions (often with several maxima), but the luminous flux, F, of luminescence accounts for a comparatively narrow region of the emission spectrum (usually with one maximum)

$$F = \int_{\lambda_1}^{\lambda_2} I(\lambda) \, d\lambda \tag{3.1}$$

where λ_1 and λ_2 are the boundaries of the luminescence spectrum.

Luminescence is also favourably distinguished by the distribution of the maximum of emission energy along the spectrum. Attention should be given to the fact that the areas of blank paper (where the luminescence has been quenched or is absent) practically do not luminesce, i.e. appear as black surfaces, the opposite of what is seen in reflected light on a conventional map. True, with excitation by ultraviolet radiation there may be a case when the excitation light rays are superimposed on the luminescence light, but this can be easily eliminated by a special filter.

Numerous lumogens are seen in the white light as white powders, which offer a possibility of creating achromatic paints—grey or white, with different luminescence spectra (Fig. 3.1).

It should be noted that conditions can be created when the colour of luminescence produced by the luminescing agent contained in the paint does not correspond to its natural colour, e.g. a brown paint can acquire a yellow or red emission spectrum upon excitation, a violet paint can acquire blue or dark

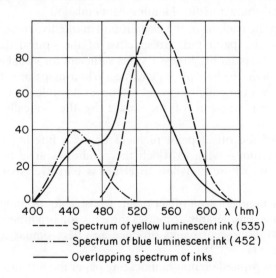

---- Spectrum of yellow luminescent ink (535)

—·—· Spectrum of blue luminescent ink (452)

——— Overlapping spectrum of inks

Figure 3.1

blue, etc. This technological feature in the application of luminescence expands the possibilities for a machine to distinguish between different quantitative data.

An extremely significant advantage of luminescence over reflected light consists in the possibility of attaining larger brightness intervals between the elements of the image that are and are not to be automatically identified.

It should also be noted that the application of luminescence provides an opportunity of using paints with radiation in the invisible region of the spectrum (infrared), which is yet another advantage of luminescence over reflected light.

Comparison of the compositions of luminous fluxes reflected from a conventional map and those radiated by a luminous map makes absolutely clear the fundamental differences between them. These differences demonstrate the indisputable advantage of luminescence, opening up extensive possibilities of applying it on maps to solve the problem of the automatic readout of both the quantitative and qualitative data.

To develop the methods for automatic readout of luminescing maps it is very important that we should know the nature of the changes in the physical properties of paints, caused by the defects and general characteristic features of offset printing.

Investigations have been performed into the most important physical properties of paints in terms of their application on maps for visual and machine reading. Listed below are the conclusions arrived at as a result of these investigations:

1. Printing with luminescent paints by the offset method provides a practically uniform luminescence brightness over the field of the background image.
2. Differences in the thickness of paint layers do not exert any significant effect on the spectral characteristic of luminescent emission.
3. The colour of the backing or the background markedly affects the emission brightness of the paint but, irrespective of the optical density of the background, the paint brightness level is sufficient for visual perception of images and their reading by a machine. When maps are designed it is nevertheless recommended to avoid whenever possible the overlapping of the luminescing elements of the image by the optically dense non-luminescing ones.
4. The colour of the offset paint pigment exerts a filtering effect on the lumogen spectrum, owing to which the conditions arise for developing luminous paints whose emission spectrum is narrower than that of the lumogen.
5. A mixture of lumogen with offset paint, whose colours complement each other, as a rule considerably decreases the emission brightness.
6. Overlapping luminescent paints produce an effect similar to that of the additive mixing (Fig. 3.2).
7. When a map is printed on non-luminescing paper the luminescence brightness of some paints is observed to decrease, as compared with brightness on conventional paper, but this decrease is of no practical significance.

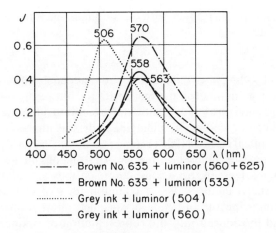

Figure 3.2

3.2. The Methods of the Latent Optical Coding of Images

3.2.1. The method of the latent optical coding of isolines

Machine-assisted reading of data represented by isolines is associated with great difficulties. Automatic identification of isolines is in principle possible, and some research and development work has been performed in this direction. However, the programmes of its implementation on a computer are extremely complicated and require long computer times. Reliability of recognition can be achieved in this case only in the absence of any noise effects from extraneous images, i.e. from the other signs of the map. In this connection it became necessary to search for such a form of isolines representation on maps which would provide for their prompt and reliable reading with the traditional image representation being preserved.

The values of isolines can be restored in different ways, primarily by the use of bergstrichs, but this is possible only in the case of two-dimensional readout and analysis of the image, which is a rather complicated and time-consuming procedure to be performed by scanning devices coupled with computers. With unidimensional readout (one line at a time) no information is actually entered into the machine by which it would be possible to judge upon the coordinate or the sign of the Δz increment, in addition to coordinates x, y. Unidimensional determination of isolines is only possible with a special marking of images. In order to make it possible to determine from one scanning line the value of z at every intersection point of the scanning element with the isoline it is sufficient to introduce into an image such features (a special code) that would allow the machine to determine only the sign of the Δz increment.

When this sign is known, the determination of coordinate z for all the intersection points of the isolines with the scanning line is reduced to algebraic summation of increments

$$z_1 = z_0 \pm \Delta z; \quad z_2 = z_1 \pm \Delta z_1; \quad \ldots \quad z_i = z_{i-1} \pm \Delta z$$

Two methods of coding have been developed that enable automatic identification of isolines in a unidimensional space with the outwardly traditional image representation being preserved (with the exception of some unsubstantial changes): one-colour coding by changing only the image brightness; multicolour latent optical coding [102, 111].

The first approach excludes the presence of extraneous images, in addition to the isolines themselves; in the second approach they are admissible. At the basis of both these methods lies a common idea of constructing a code combination following the principle of specified alternation of features (or properties) and introduced into the representation of isolines that are then decoded in the course of reading (with the aim of determining the increment sign) as the scanning element of the reading head passes from one isoline to another. Three-feature coding is regarded as optimal. For more complex, e.g. topographic, maps four-feature coding may be applied.

Figure 3.3 shows (1, 2, 3) the first, second, and third coded features, respectively (they will henceforth be designated as signals). Let us assume that in all the cases of signals alternation the increment sign is positive, i.e. in this case, when going along the lines of the profile, there is as it were an upward rise, while with the 1–3–2–1 alternation of signals the sign is negative. When the first method is used the coding signals are characterized by the gradations of isoline colour brightness, e.g., signal 1 can be represented by isolines of light-grey colour, signal 2 medium grey colour, and signal 3 dark grey colour (any other colour can be taken instead of grey). In the second method the coding signals are actualized by multispectral luminescing fillers, introduced into the paint of

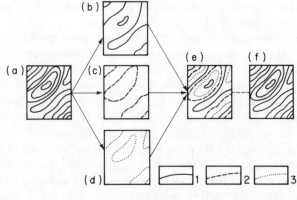

Figure 3.3

a certain colour, e.g. grey or brown [120]. Signal 1 can be represented by isolines having a green emission spectrum upon the excitation of luminescence (an ultraviolet source is desirable), signal 2 by isolines having a red spectrum, and signal 3 a blue spectrum. Under normal illumination the presence of a luminescing filler in the image of isolines does not manifest itself.

Let us now consider in greater detail the second coding method. As already noted, it allows the presence of any other substantive elements of the map, which do not have any pronounced effect on identification reliability. This is possible because all the elements of the map, except the isolines, are applied (printed) with a non-luminescing paint. The changes observed in the brightness of isoline luminescence (where the isolines overlap the non-luminescing elements of the image) affect the amplitude of the signal recorded by the reading device. But, owing to the fact that the luminescence of paints has a sufficiently large spectral spacing (the spectra do not actually overlap), the signal is completely filtered off, without any noise, with the help of optical filters alone. This allows a free choice of the necessary discrimination threshold for the signal, ensuring the functioning of the system with the minimal signal amplitude. As shown by experience, selection of isolines and the optical filtration of their signals are performed with sufficient reliability.

A map with such latent optical code should be edited and published with due regard to the following requirements. The original copy of the map of isolines (Fig. 3-3a) is broken up into three smooth-delineation maps (Fig. 3.3b,c,d) so that when they are matched in the course of being printed with luminous paints (Fig. 3.3e) a specified sequence of luminous signals alternation should be provided when the map is read out by a special device (when the map is devised from the data of field surveys it is advisable to prepare the separate originals simultaneously). Such a sequence of signals can be provided by isolines printed with three paints which, when excited, produce different emission spectra. It is expedient to assign the spectral maximums of each paint with a sufficiently large spectral spacing, e.g. for a ternary sequence of signals, in the blue, green, and red spectral regions. If a map of isolines with such a code is scanned as shown in Fig. 3.3, the positive increments will arise with the blue–red–green–blue alternation of signals, and the negative ones with the blue–green–red–blue alternation.

The composition of luminescing paints is based on achromatic offset paints and organic luminors, preferably without their own pigmentation, i.e. colourless or white. The following luminors can be recommended for such paints: yellow (560), green (517), blue–violet (452), red (640), green (505), and some others. When the maps are printed it is first necessary to apply all the other substantive elements of the map (e.g. roads, rivers, or inscriptions), with conventional offset paints and then draw the isolines with a luminescent paint.

For complex images one can use a code combination with four or five paints having widely spaced spectral characteristics. This may be necessary to show half-interval contours on a topographic map.

3.2.2. The method of obtaining multicoloured maps with a latent stereoscopic image

The proposed method is based on the principle of colour separation of the stereoscopic pair of conjugated images using the properties of luminescence [103]. This method makes it possible to compose a map that can be used both as a conventional multicoloured (e.g. topographic) and a stereoscopic map when irradiated by ultraviolet light. In the latter case the image is viewed through anaglyphic spectacles. The stereoscopic model is visually perceived in natural lighting and more clearly in darkened conditions. This effect is achieved by applying (printing) both conjugated images with special luminescent paints. One of the images is printed with invisible (in natural lighting) luminescent paint based on organic luminophor luminescing in a short spectral region (e.g. blue–violet luminor 452). The other image is applied with visible luminescent paints of several colours (corresponding to the colours of the shaded elements of the map), composed on the basis of one luminophor with the medium-emission spectrum region (e.g. yellow–green luminor 535), whose luminescence colours complement that of the invisible luminescent paint. The luminescence is excited by ultraviolet rays ($\lambda = 365$ nm). The stereoscopic image is viewed through spectacles with two light filters whose spectral transmission maxima correspond to the maxima of the emission spectra of the luminors contained in the paint. The stereoscopic model is constructed following the principle of the additive method of colour anaglyphs. The printing of such maps includes the following operations:

1. Applying the background non-luminescing visible elements of the map image with conventional paints.
2. Applying the visible conjugated shaded image of the stereogram with colours corresponding to this image (with certain corrections).
3. Applying the other conjugated shaded image with invisible luminescent paint.

For the stereoscopic image to be perceived as light-colour (luminous) lines against a dark background, maps are printed on non-luminescing cartographic paper. For the stereoscopic image to be perceived as dark lines against a light-colour background, maps are printed on special luminescent paper with white-colour luminescence.

Luminescent paper giving white luminescence is manufactured by treating ordinary paper with two luminors entering into the composition of paints with which the stereoscopic image is applied, or by using some other luminors with a similar spectrum but mutually complementing one another by the colour of luminescence (e.g. luminors having a blue and yellow luminescence). The following conditions should be met in this case:

1. For the images to be readily separated the emission spectra of the luminors should not extensively overlap one another. After the paper has been

treated with such luminors its emission spectrum should have two clearly defined maxima.

2. The colour of luminescence emitted by luminors and their concentration in solution prepared for the treatment of paper should be such that when the paper is viewed (after treatment) under ultraviolet rays the background is perceived as white, both with and without the light filters.

3. The quantity of luminor included in the paint should allow the image (applied with this paint), when perceived through a corresponding light filter under ultraviolet rays, to merge with the background, e.g. in the case of a luminescent base the image with blue luminescence is viewed through a blue light filter and with yellow luminescence through a yellow–orange filter.

4. When the invisible image is viewed on white luminescing paper through a filter whose transmission spectrum maximum corresponds to that of the emission spectrum produced by the luminescence of the invisible image paint, it is necessary that the luminescence brightness of the image should be the same as that of the background (in our example, when viewed through a blue filter). When the same image is viewed through a filter whose transmission spectrum maximum does not coincide with the emission spectrum maximum for the luminescence of the paint with which the invisible image has been applied (i.e. does not transmit or only partially transmits its radiation), it is necessary that the invisible image should be perceived as grey lines (in our example, through a yellow–orange filter). The opposite must occur when a visible image is viewed.

Only when all the above conditions are met can a good stereoscopic effect be achieved. As shown by experience, the most distinctly perceived stereoscopic image is attained when the stereogram is applied on paper having a white colour of luminescence. In this case the stereoscopic effect is observed under daylight illumination as well. The stereoscopic image is perceived as grey lines against a white background.

When visible paints are manufactured, one and the same luminor is introduced, irrespective of the colour sought. Note that emission spectra of visible luminescent paints are somewhat narrower than the luminor spectrum, with the maxima shifted in either direction, depending on the filtering effect of the offset paint. This, however, does not interfere with the construction of a stereoscopic model, provided the light filter completely suppresses (cuts off) the whole region of the emission spectrum of the luminor complementing the paint by its luminescence colour, which has been confirmed by actual experience.

As already mentioned, a visible shaded image is subjected to colour correction, i.e. a certain change in the colour of individual images relative to the conventionally used ones. For instance, on topographic maps all the elements of blue images are replaced with bluish green, and of black images are replaced with grey. This has to be done because the luminor introduced into offset paints has an emission spectrum differing (sometimes greatly) from the reflection spectra of some paints, which results in a decrease in the brightness of the luminor luminescence.

The invisible image of a map is applied with a paint composed of white pigment, a luminor without inherent pigmentation, drying oil, and a softening paste. In achromatic luminescent paint compositions the relative energy distribution along the luminescence spectrum does not actually change as compared with the spectrum of the luminor, but its luminescence brightness is noticeably lower.

For such stereoscopic models to be read by a machine, i.e. for the construction, based on them, of the same digital models as those obtained by the preceding method, a special device has to be developed. The stereogram image in it can be separated with the help of the same light filters as those used for visual perception.

The method can find application for educational demonstration maps and various special-purpose maps (maps for air and space navigation, etc.).

3.2.3. The method of latent optical coding of data based on luminescence persistence

The use of the properties of luminescence persistence for the latent map data coding by a computer expands the possibilities of information retrieval and improves the reliability of image selection both by the emission spectra and the persistence of luminescence. This, in turn, makes it possible to increase the information capacity of maps. Since only few luminescing substances manifest prolonged persistence (as a rule, these are crystallophosphors), it becomes possible with their help to create additional images, distinguishable from the other images on the map by their longer persistence. The method proposed is as follows.

Luminous compositions of crystallophosphors are selected, whose persistence is of longer duration than that of organic luminescing substances. The law of luminescence attenuating for the latter is expressed by a well-known exponential dependence

$$I = I_0 e^{-(t/\tau)} \tag{3.2}$$

where I is the luminescence intensity at the moment of time t; I_0 is the luminescence intensity at the end of excitation; τ is the average duration of excited state (the time within which the intensity of luminescence becomes e times lower). As different organic substances have the excited state duration varying within $\tau \sim 10^{-8}$–10^{-9} seconds, for luminous compositions made up of crystallophosphors it must exceed 10^{-8} seconds, for example, for FKP-03-01, FK-101, and other composite phosphors. Paints (inks) are prepared from crystallophosphors with prolonged persistence. Casein, organic resins, and other substances can be used as binding agents. The concentration of phosphors in the paint has to be high enough to assure the necessary luminescence intensity, I_0, at the moment of recording the optical signal by a photoelectron receiver within the time interval Δt after the cessation of excitation. The paint thus prepared is used to apply the necessary graphical data on a map. The data can be represented by all kinds of cartographic symbols, such as a code of map notations for data search or a line limiting the area of the map to be read by a scanning device.

Images can be applied manually or with the help of an automatic graph plotter having several writing pens charged with these paints, and, possibly, also the means of printing (e.g. by the silk-stencilled method).

In the course of data search and readout the image applied with a paint containing a long-persistence composite phosphor is irradiated by ultraviolet rays. Upon the cessation of luminescence excitation the persisting signal is recorded by a photoelectronic receiver.

For automatic readout and identification of symbols, proceeding from the attenuation (persistence) curve, it is expedient to use electromechanical scanning devices. The light spot in these devices is formed by an ultraviolet illuminator, with quartz condenser and the scanning element spaced in the direction of scanning at such a distance and time interval as to allow a discrete recording of the attenuation curve in several points. Discrete signals of the curve in the form of a series of codes enter the comparison block designed to identify the symbols (signs) applied with a paint containing luminors with a specified persistence constant.

3.3. Comparison of Paints and Inks Used for Normalized Maps

3.3.1. Luminescent offset paint *

The paint is designed for the application (printing) of a luminescing image in the form of a latent optical code that is decoded by a reading device with the help of ultraviolet irradiation and an electronic processing block. The paint is distinguished by a heightened yield (intensity) of luminescence under the action of ultraviolet rays and by its stability in storage. The paint contains a binder composed of polycondensates rosin with maleic anhydride, phthalic anhydride with pentaerythritol, and polymerized flax-seed oil in the ratio of $1:1:3$ (vehicle 11003), includes an asphalt fraction with a boiling point at 350–380°C (MP-1) and also a solution of maleated rosin pentaerythritolate in a mixture of flax-seed and transformer oils in the ratio of $1:1:2$ (varnish 6-03), solution of phenolformaldehyde resin in flax-seed oil in the ratio of $1:1$ (varnish 6-04) and polyethyl with a molecular weight of 8–12 thousand (paste PP), with the following component ratios, parts by weight:

Luminophor	380–400
Pigment	100–125

---

* Inventor's certificate N 852921, *Bull. izobr.*, N 29, 1981.

Mixture of the polycondensates of rosin with maleic anhydride, phthalic anhydride with pentaerythritol, and polymerized flax-seed oil	360–390
Asphalt fraction with b.p. 350–380°C	25–45
Solution of maleated rosin pentaerythritolate in a mixture of flax-seed and transformer oils	15–25
Solution of phenolformaldehyde resin in flax-seed oil	10–20
Polyethylene	20–30
Naphthenate–cobalt drier	10–20

3.3.2. Non-luminescent offset paint*

The paint is designed for the application (printing) of a non-luminescing image intended for visual perception only. Characteristic features, distinguishing it from conventional offset paints, are its negligibly small luminescence capacity, practical opacity with respect to ultraviolet radiation, and heightened abrasive resistance of map prints. The binder in the paint is synthetic glyptal drying oil (SGDO) and the alkyd is natural flax-seed drying oil, with the following component ratios, parts by weight:

Titanium dioxide	580–700
Drier	50–60
Synthetic glyptal drying oil	140–190
Natural flax-seed drying oil	90–120
Asphalt fraction with b.p. 350–380°C (MP-1)	20–50

3.3.3. Luminescent ink†

The main function of the ink is to normalize images designed for machine processing. This ink can be used to mark traditional cartographic materials, and to apply normalized images on aerospace photographs and other materials designed for subsequent automatic processing.

Distinguishing features of the ink are its heightened yield of luminescence under the action of ultraviolet rays, high water resistance, and stability in storage. The ink contains as an antiseptic a mixture of formalin, borax, and phenol at 1:1:0.6 weight ratio. It also contains an acid or basic dye, ammonia (25 per cent) and casein, with the following component ratios, parts by weight:

Luminophor	22–28
Antiseptic	36–44
Dye	1–9
Casein	100–130
Ammonia (25% water solution)	5–6

---

* Inventor's certificate N 836063, *Bull. izobr.* N 21, 1981.
† Inventor's certificate N 852925, *Bull. izobr.* N 29, 1981.

Ethylene glycol   40–50
Water   740–790

3.4. Conclusions

Summing up what has been presented in this chapter we should like to emphasize the principal advantages of luminophors (organic and inorganic composite phosphors) and the luminescent paints and inks based on them and used to compose normalized cartographic images for automatic read-out.

1. Luminophor-based invisible paints and paints with diverging spectral characteristics (i.e. those of the reflection spectrum in daylight illumination and of the luminescence spectrum in ultraviolet rays) make it possible to produce normalized maps of high information capacity by means of constructing additional images containing a latent optical code designed to be read by a machine and to compose multicoloured stereoscopic maps perceived visually in the daytime and in ultraviolet rays at night.

2. Paints and inks whose compositions are based on luminophors have such optical properties as:
   (a) narrow spectral band of luminescence radiation;
   (b) luminescence of some compositions in the near infrared zone of the spectrum;
   (c) different persistence constants for different compositions based on inorganic luminophors;
   (d) relatively large brightness intervals between the luminescing elements and the background;
   (e) luminescence scattering (diffusion), irrespective of the direction of excitation radiation.

All this makes it possible to devise on a map a great variety of optical codes and graphical symbols reliably indentified (selected) by a computer (see section 4.1.2).

CHAPTER 4
Automatic Reading of Maps

Automatic reading of maps is understood as the conversion of a cartographic model into its digital analogue—a digital cartographic model, represented in machine codes in a selectively (with respect to the qualitative and quantitative characteristics of objects) obtained form. In other words, it comprises the readout, interpretation, and recording of cartographic data on a machine data medium in digital form with the help of automatic reading devices and a computer.

Depending on the data representation method used on a map, and the nature of their spatial distribution, the digital model can acquire different forms. Among the basic forms are the following:

1. Continuous data, represented by the isoline method. They can be converted into a regular and semi-regular digital matrix showing the spatial coordinates of the relief surface, a geophysical field, and other objects, or into a vectorial form, following the traces of isolines, i.e. into vectors between the two successively perceived control points on the isolines.
2. Continuous–discrete data, represented by the methods of qualitative background, areals, cartograms, including the methods with raster discretization. They can be presented in a vectorial form by Cartesian coordinates along the contour boundaries together with a code of a qualitative and/or quantitative parameter, or in 'cellular' form, i.e. in the form of a regular matrix consisting of digital raster points.
3. Discrete data, represented by the methods of symbols, localized diagrammatic maps, digital code signs, etc. They can be represented by the coordinates of their centres and a code of their qualitative and/or quantitative characteristic.

All the methods of cartographic image conversion into a digital code, dealt with in this chapter, are based on data readout and input into a computer by means of scanning devices. This is associated with the fact that the scanning method assures complete automation of computer data input. It should be added that scanners provide for a sufficiently high spatial and frequency resolution and are also relatively inexpensive, which seems to have been the reason for their worldwide usage [5, 18, 27, 29, 59, 79, 89, 107, 130, 134, 142, 153].

When scanners are used as input devices only (universal scanners) all the functions of converting an image into a digital model, including the selection of data by their qualitative and/or quantitative parameters (i.e. identification of symbols), are assigned to a computer. This imposes special requirements on the development of specialized software. It becomes practically impossible to meet these requirements if one takes into consideration the multifarious forms of traditional cartographic images. However, with the introduction of unified principles of cartographic data normalization it has been proved possible to develop a software package enabling automatic conversion into a digital model [111, 120, 131, 143–145].

The methods of automatic tracing of lines or contours, i.e. of bringing the points obtained by scanning into an ordered sequence relative to the assigned vector, which are discussed in this chapter, can be applied to any kinds of images, both normalized and traditional. To ensure reliable automatic identification of symbols on the map different procedures of partial or complete normalization of images are used. On the newly designed special-purpose maps it is expedient to present their thematic content in a fully normalized form. The already existing maps are digitized using the procedures of partial normalization, necessitating small expenditures of labour. Identification of symbols and conversion into a digital model are performed on fully normalized maps faster and more reliably than on the partially normalized ones. The total normalization of a map is, however, worthwhile mainly in updating the old and designing completely new maps.

Conversion of a cartographic image into digital form is necessary for the solving of numerous research and production problems.

Research, investigation, prospecting, design, and other kinds of work can be performed in an automatic mode directly based on normalized maps. In a number of cases it may prove reasonable to start with developing a bank of digital data obtained from normalized maps and then solve problems with the help of digital models.

4.1. The Methods of Selective Data Readout by the Colour Characteristics of Cartographic Images

4.1.1. The problem of identifying map symbols by the colour characteristics of images

The most widely practised representation method for special-purpose and thematic maps is the method of qualitative background based on using a colour hue of varying saturation and brightness with the inclusion of additional denotations in the form of shadings, indices, etc. (mostly of black colour). For such representation the colour characteristics should be regarded as the basic distinguishing features of symbols. The principle of identification by colour characteristics must obviously lies at the basis of designing the devices for automatic readout and identification of symbols.

By now, extensive research and development work has been devoted to colour identification, both for the spectral colours and those of paints and transparent colour media. It should be noted that colour identification capacity differs significantly, depending on the form of colour representation. Identification of the colour of painted surfaces is much more complex and less reliable than in the case of transparent colour media. Numerous specialists have studied the questions related to the identification of the hues of paints on opaque substrates, mainly as applied to the reading of coloured charts and diagrams.

These problems arise because of the original material itself (i.e. its form of data representation) and are also caused by the fact that the reading and identification of colours have to be performed in reflected light in dynamics. Identification reliability in this case primarily depends on the following parameters: texture and spectral characteristics of the material (data medium); degree of uniformity of coloration hue, saturation, and brightness; image quality. In contrast to charts and diagrams, where lines are usually thicker and colours more homogeneous, colour identification on maps is much more complicated. Symbols on a map are represented not by a solid filling of a certain uniform colour but by different kinds of screens differing both in frequency and shape, and, what is most essential, differing in hue on one and the same symbol. For instance green colour is not infrequently produced by the overlapping of blue and yellow screens, violet is produced by blue and red screens, etc. (with the lines of different frequency and thickness). Solid colour fillings account for a comparatively small percentage of map surface. The overlapping of screens results in subtractive and additive mixing of colours, governed by different laws.

All this makes it necessary to perform two-dimensional analysis of the screen structure or to use a sufficiently large scanning element for the points (sites) of overlapping and non-overlapping colours to be summed into a single unified luminous flux, providing the homogeneity of colour and its distinction from the colours of other symbols. The choice of the scanning element dimensions is determined by the parameters of screens, the form of their overlapping, the character of the image, and the quality of printing. An increase in the size of the scanning element will obviously decrease the accuracy of readout.

Identification of map symbols by their colours is a specific case in the overall problem of colour identification. What is the optimal way of solving this specific problem? Since, when designing the map image and reading it, we proceed in the general sense from colour sensitivity and the discrimination threshold of the human eye, and the search for a given symbol on the map and its recognition are made by comparison with the symbol of the legend (the standard), it quite naturally follows that a cybernetic model of visual colour perception should be constructed.

For the identification of colour symbols on a map the following method, developed by the author, is proposed. The principle of amplitude selection can be used to identify a limited number of colour hues. In accordance with this principle it is expedient to choose in advance the colours of symbols with the necessary spectral characteristics. It is only possible to achieve a more or less reliable colour identification, based on this principle, on a specially prepared map with carefully selected colours.

4.1.2. Selective readout of multicoloured images with the help of a single-channel device

Let us examine the principle of selective readout with one light-sensitive element, as exemplified by a multicoloured luminescent image. The method is founded on double selection: optical, by means of colour filters; and electronic, by means of amplitude selectors (or discriminators).

Primary selection with the help of colour filters has the functions of suppressing the redundant luminescence data and isolating a maximum of legitimate data.

In selective readout of qualitative data by the assigned valid signal level it is first necessary to establish the value of the maximum level of the noise signal, resulting from the overlaps in the spectra of redundant luminescent images being incompletely cut off by the colour filters.

Let us assume that the whole luminescent image has practically the same brightness. It is necessary to determine the ratio of the valid signal level to the noise level. To determine this quantity one first of all has to estimate it for all cases of the mutual overlapping of spectra and then proceed from its minimum value.

The value of luminous flux (the valid signal of luminescence, q) falling onto a photocathode, can be expressed as

$$F_q(v) = S\omega u \int_{\lambda_{q_1}}^{\lambda_{q_2}} I_q(\lambda)\tau(\lambda)\,d\lambda \tag{4.1}$$

where S is the area of the luminous spot (scanning element), ω is the solid angle; u is the spectral transmission factor of the optical system; $I_q(\lambda)$ is the spectral distribution of the luminescent paint emission brightness; $\tau(\lambda)$ is the spectral transmission of colour filter; and λ_{q_1} and λ_{q_2} are the boundaries of the spectral region of luminescent emission.

The saturation photocurrent (of the valid signal) induced in the photocathode circuit by flux $F_q(v)$ will be equal to

$$i_q(v) = S\omega u\varepsilon \int_0^{\lambda_{q_b}} I_q(\lambda)\tau(\lambda)\gamma_0(\lambda)\,d\lambda \tag{4.2}$$

where ε is the maximum spectral sensitivity of the photocathode; $\gamma_0(\lambda)$ is the relative limit of the photocathode spectral sensitivity; and λ_b is the photoelectric threshold of the photocathode.

Similarly the value of the noise signal caused by luminescent emission r,

$$i_r(n) = S\omega u\varepsilon \int_0^{\lambda_b} I_r(\lambda)\tau(\lambda)\gamma_0(\lambda)\,d\lambda \tag{4.3}$$

The ratio of the valid signal to the noise signal will be expressed as

$$K = \frac{i_q(v)}{i_r(n)} = \frac{\int_{\lambda_0}^{\lambda_b} I_q(\lambda)\tau(\lambda)\gamma_0(\lambda)\,d\lambda}{\int_{\lambda_0}^{\lambda_b} I_r(\lambda)\tau(\lambda)\gamma_0(\lambda)\,d\lambda} \tag{4.4}$$

Since the $\tau(\lambda)$ and $\gamma_0(\lambda)$ quantities appear under the integral sign in both the numerator and the denominator, for their numerical values to be obtained it is sufficient to assign them in relative units. From a series of obtained K values K_{min} is chosen, and from it the noise level limit established. With weak luminous fluxes one should take into account the dark currents of photosensors.

We shall now give an example of calculating the value of K for orange and light green luminescent paints, assuming the orange paint emission to be the legitimate signal.

Figure 4.1 shows the curves: of the spectral sensitivity of FEU-22 device, $\gamma(\lambda)$; of the spectral transmission of OS-II colour filter, $\tau(\lambda)$, of the emission spectrum of orange paint, $I_q(\lambda)$, of the emission spectrum of light green paint, $I_r(\lambda)$, and the curves obtained as the result of integration.

Having integrated expression (4.4) we shall get in the nominator the area, formed by the $I_q(\lambda)\tau(\lambda)\gamma_0(\lambda)$ curve and the λ axis, and in the denominator the area bounded by the $I_r(\lambda)\tau(\lambda)\gamma_0(\lambda)$ curve and the λ axis. Having then divided these areas, we shall obtain $K \approx 10$. The value of K shows that the valid signal is several times higher than the noise signal despite the comparatively strong overlapping of the luminescent paints emission spectra. This example shows that, with the resolving power of present-day photosensors and other elements of readout systems, high degrees of emission spectra overlapping are admissible.

The function of the discriminator is to isolate the pulses of the legitimate signal, whose amplitudes exceed the assigned voltage level below which lie the

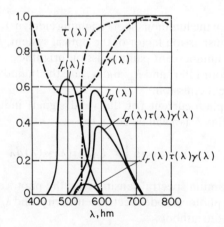

Figure 4.1

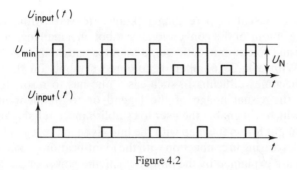

Figure 4.2

noise signals. Selection threshold is chosen above the highest amplitude of pulses caused by the superfluous luminescent image (resulting from the overlapping emission spectra being incompletely cut off by colour filters). Selection threshold U_{min} depends on the bias voltage at the discriminator input (Fig. 4.2).

4.1.3. Identification of symbols by their colour characteristics

In terms of modern chromatics, coloration of any reflected luminous flux can be presented as the result of the mixing of three basic colours X, Y, Z in certain proportions x', y', z'. This can be expressed by colour equation

$$F = x'X + y'Y + z'Z \qquad (4.5)$$

Let us assume that intensity distribution in the spectrum of a given radiation is assigned by $I(\lambda)$ function, and the distribution of the intensities of three basic radiations are represented by curves $\bar{x}(\lambda)$, $\bar{y}(\lambda)$, $\bar{z}(\lambda)$, where $\bar{x}$, $\bar{y}$, $\bar{z}$ are the specific colour coordinates. Any luminous flux reflected from a coloured surface, including that of a map, has a continuous emission spectrum. The radiation composition of such a spectrum can be expressed as the product of spectral intensity $I(\lambda)$ of a source by spectral reflectivity $p(\lambda)$.

Using special colorimeters one can determine any colour on a map and express it in the coordinate form:

$$x' = \int \bar{x}(\lambda) I_v(\lambda) \rho(\lambda) \, d\lambda;$$

$$y' = \int \bar{y}(\lambda) I_v(\lambda) \rho(\lambda) \, d\lambda; \qquad (4.6)$$

$$z' = \int \bar{z}(\lambda) I_v(\lambda) \rho(\lambda) \, d\lambda;$$

where $I_v(\lambda)$ is the spectral intensity of the radiation of the source, v; and $\rho(\lambda)$ is the spectral reflectivity of paper (map surface) for radiation with the wavelength of λ.

Let us first consider the conventional process of the visual identification of a colour symbol on a map.

The process of visual map reading is known to start with viewing and memorizing the colour of the conventional symbol on a map legend and then, based on the visual colour image, the same colour is searched for and identified on the map. With a large number of coloured symbols on the map their identification becomes difficult. In such cases one has to resort to repeated restoration of the visual image of the legend or to frequent momentary comparisons, which will enable the user to establish more reliably the identity of the colour sought for on the map with the hue given in the legend.

The difficulties arising in connection with the identification of colour symbols of similar hues are explained by the limited resolving power of the human eye and, to a great extent, by its adaptation to light and colour, as well as a certain degree of vision persistence. It is possible to exclude the effect of errors associated with the peculiarities of human visual apparatus only when the given and the identified colour hue are situated in a similar environment and visualized simultaneously. Such visualization and comparison of coloured symbols are often used by cartographers when copying colours in the course of colouring a map according to the assigned scale. The principle of comparing two optical fields at the same moment of time is widely applied in visual photometers that assure a high accuracy of measurements. All this gives us grounds for believing this method to be the most accurate, and that it should be used as the basis of developing new readout systems. Considering the above-mentioned characteristic features of visual perception, it is equally important for us to take into account the functioning of the colour-perceiving elements of the human eye. In choosing the photosensing elements for readout systems by their spectral sensitivity one should proceed from the tricomponent theory of colour vision and the system of tricolour coordinates existing in chromatics. It is therefore necessary to take three photosensing elements with the maxima of spectral sensitivity relative to the blue, green, and red spectral regions and provided with colour filters having the corresponding spectral transmissivities.

To meet the requirement of perceiving simultaneously the assigned colour taken from the legend (henceforth referred to as the standard) and the colour to be identified on the map, coinciding with the standard, it is necessary to create two absolutely identical receivers: one to perceive the standard colour, and the other to search for an identical colour on the map (using the method of simultaneous perception and comparison of the signals coming from the standard and the background of the map). It is also necessary for the standard and the map to be exposed to the same degree of illumination.

Proceeding from the above conditions, it is proposed that a device, whose block diagram is shown in Fig. 4.3 [98, 100], should be created.

The device contains an illuminant (1), a prism (2) that divides the luminous flux into two fluxes focused by condensers (3) on the conventional sign of the legend (the standard) (4) and the map (5), respectively. Radiations from the conventional symbol of the legend and from the map, each directed via their channel, are recorded by two identical sensors consisting of a lens (6), a trihedral prism (or semitransparent mirrors) (7), dividing the luminous flux

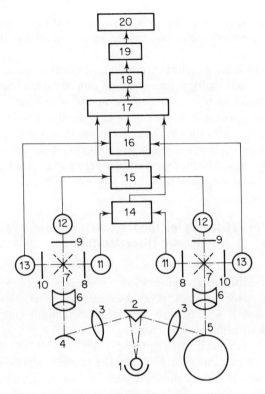

Figure 4.3

into three intensive fluxes directed correspondingly via the red (8), green (9), and blue (10) colour filters to three photometric tubes (11, 12, and 13). Signals from the outputs of the three identical photomultiplier tubes enter comparator units (14, 15, and 16), consisting of NOT gates and amplifier-forming stages. The outputs of comparator units are connected to an AND gate (17), whose signal through a former (18) and an amplifier (19) enters the data processing block (20).

The signal at the output of AND gate (17) arises when the radiation from the map at the moment of its scanning coincides in all the three colour characteristics with the colour of the assigned conventional symbol of the legend.

This device for the readout of qualitative data by their colour characteristics can be used for traditional maps with the colour qualitative background.

It is possible to create an additional block for this device so as to eliminate the negative effect of the shaded elements of the cartographic base on the accuracy of readout.

On the basis of the tricolour system of colour coordinates one can construct a device for discrete identification. The device can contain a light sensor, a memory block with three registers, a comparator, and an AND gate. In this device the readout is performed sequentially and discretely, i.e. first to be inserted into the memory are the colour characteristics of the standard (the assigned legend symbol), after which the map is read by means of linear scanning.

This principle of colour identification can, in general, be realized with the help of computers and multipurpose readout devices actualizing the tricolour input. The problem can also be solved by means of computer programming. True, this discrete principle is less reliable than the principle of analogue modelling. It may be said that a photoelectronic system modelling the visual process of colour perception is capable of producing results whose reliability is similar to that of human vision, especially if a perceptron simulating the human eye retina is used as the reading head.

4.2. Techniques of Identifying the Data Represented by the Method of Raster Discretization

As already noted, the method of raster discretization, as distinct from the traditional methods of data representation, makes it possible to identify data in a unidimensional space (the same property is also characteristic of isolines with a latent optical code). This feature of MRD substantially simplifies the procedure of data identification by a machine.

Identification can be performed both by specialized blocks and a computer. When identifying the raster symbols by MRD a scanning device may operate in the triggered scanning mode, distinguished by high speed and simplicity of design. Most of the scanners used today are devices with triggered scanning.

We shall now consider some methods of identification with the help of a computer (when the reading device performs only the input functions) and specialized devices.

4.2.1. Identification of data by the parameters and combinations of linear rasters with the help of a computer

Let us first examine the principles of identifying symbols constructed by means of MRD, when all the identification procedures are assigned to a computer, and the readout device has the input functions only [120].

Identification reliability is to a great extent determined by how optimally the ratios of line thicknesses to the scanning head size and the discrimination threshold have been chosen. We suggest the following criteria for the optimum variant:

1. the threshold of discrimination between the points representing the background and the image (for lines of the same brightness) by their quantization level must correspond to such a position of the scanning element when 50 per cent of its area is occupied by the image;
2. the minimum thickness must be determined by two points (elements);
3. the differences in line thicknesses must amount to two or three points.

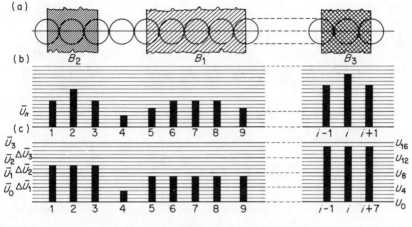

Figure 4.4

Identification of symbols by the intervals between lines is similar to the identification by the thickness of raster lines, the only difference being that the ratio of signals by their quantization levels becomes diametrically opposite.

Identification of symbols by the line brightness is performed based on the codes corresponding to the assigned brightness quantization levels for signals. Such identification is only possible after preliminary preparation of a massif by bringing the adjacent points (whose quantization level exceeds the threshold of discrimination between the background and the image) to the same level.

Figure 4.4a shows the mutual positions of the scanning head and the raster line in the course of scanning. Figure 4.4b shows the quantized signals arising at each discretely recorded position of the scanning element. Figure 4.4c shows the signals transformed as a result of the preliminary preparation of the massif. Prior to transformation, the points are separated into the background and the image. The discrimination threshold is assigned higher than the level of the background point. The idea of transformation consists in bringing all the adjacent (neighbouring) points to the same quantization level, assigned by the maximum value of the whole series of adjacent points. For instance, for the first line it will be point 2 (see Fig. 4.4b); consequently, points 1 and 3 have to be brought to this level (see Fig. 4.4c). It is possible to bring the points to the same level by the mean value of the codes of points. In this case though, the identification will be less reliable.

Let us now assume that there are three line brightnesses gradations: B_1, B_2, B_3. These lines have been read out with sixteen quantization levels (see Fig. 4-4c).

It is first necessary to define the boundaries separating the quantization levels with respect to the brightness gradations. This is done by dividing the quantization levels located above the threshold, $\bar{U}_t$, into a certain number of brightness gradations (in this example into three levels, $\bar{U}_1$, $\bar{U}_2$, $\bar{U}_3$). This results in all the points whose quantization levels correspond to brightness B_1 falling within the

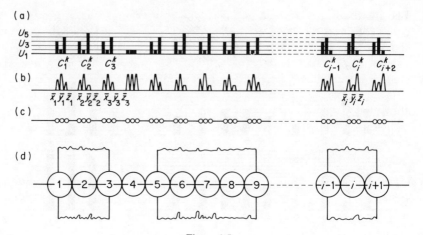

Figure 4.5

$U_4 \div U_8$ interval, the points with brightness B_2 within the $U_8 \div U_{12}$ interval, and the points with brightness B_3 within the $U_{12} \div U_{16}$ interval.

To single out a specific brightness gradation it is sufficient to assign machine codes corresponding to a given interval of quantization levels. Line thicknesses are identified by the number of successive neighbouring points with the same quantization level code corresponding to a given line brightness (i.e. points having no points with another brightness code in between).

Identification of lines by their colour, as well as by combinations of several raster parameters, should be performed by placing a reference symbol (the standard) into computer memory and then comparing each time the readout image with the standard.

Figure 4.5d shows an example of three-element raster lines, two of which are green elements (reference numbers 1 and 2) of differing thickness, and one is red (reference number 3). Numbers 1, 2, 3, ..., $n + 1$ denote different positions of the scanning element in the course of raster lines scanning.

Figure 4.5c shows the points recorded by three channels. Each discretely fixed position of the scanning element, 1, 2, 3 ..., has three corresponding points. Each of them represents its own colour channel. All three points together represent the integral colour characteristic in the form of three colour coordinates, x_i, y_i, and z_i, within the area covered by the scanning head. The ratios between the colour values of signals at the scanning head outlet are shown in Fig. 4.5b.

Figure 4.5a shows the ratios between quantized signals after their transformation. Each quantization level, $U_1, U_2, \ldots, U_i$, has a corresponding digital code stored in computer memory.

Identification by the standard (the symbol in the map legend) prerecorded in machine memory can be performed with the help of different methods. The algorithms of some of these methods will be briefly examined below. All the methods are founded on two basic principles:

1. The standards are read out and placed in computer memory for the whole combination of raster lines (constituting a group) forming a conventional sign (or signs) of the map legend, and then a given symbol is identified by comparing a raster line combination readout from the map with the standard stored in computer memory.
2. The standards are read out and placed in computer memory separately for the individual parameters of all the raster lines used on a map, and then the individual parameters denoting certain elements of complex phenomena are read out and identified, to be represented in the legend in the form of a combination of several raster line parameters.

The following three identification methods based on the first principle can be proposed.

The first method

The problem is formulated as follows: all the contours included in the map are to be read, which implies the identification (discernment) of the coordinates of contour boundaries, enclosing the areas within which certain objects or their individual characteristics occur in accordance with the symbols denoting them on the map and then 'packing' these coordinates in a certain logical sequence in the computer storage medium. The map-reading procedure by this method comprises the following consecutive operations.

All the notations in the map legend are read out, coded and recorded as standards in the computer memory.

The number of lines read out in each symbol is in this case exactly equal to the number of lines in the group. At the moment when the scanning element confronts the boundary line of a contour (usually of achromatic colour), distinguished by high brightness quantization levels as the image of the highest optical density, the coordinates of the line are recorded in the computer memory. As the scanning element passes the first group of lines (constituting a symbol) after the contour boundary its data are compared with all the standards, i.e. the legend notations stored in the computer memory as codes. It is thereby ascertained which standard is the most similar to the group of lines read out. If such a notation has been found the coordinates obtained when the contour boundary was traversed are assigned the code of the given standard. The same code is assigned to the coordinates of the contour boundary when the scanning element leaves the contour. If for some reason or other (e.g. because of image defects) the contour is left unidentified, its coordinates are sent to a separate memory block for subsequent additional identification by tracing. The process of reading with simultaneous identification is followed by the procedure of tracing the points on the inner edge of the contour boundary line, aimed at obtaining an ordered sequence of coordinates for a single-thickness (consisting of one series of points) contour boundary line.

The second method

The objective is to read out selectively the contours relating to only one given object or phenomenon.

In this method only one symbol of the legend, relating to the specified object, is read out from the map, introduced into a computer, and formed as the standard. The latter is formed by repeatedly scanning one group representing the symbol. All the code sequences arising in this case are recorded in the memory block of the computer. Repeated scanning is necessitated by the ambiguity of code point sequences, caused by a differing position of the scanning element relative to raster lines at every new line of scanning (due to the device errors), as well as by image defects. In the first method each of the groups of raster lines read out has to be compared with all the symbols in the legend, whereas in this method the comparison is made with only one standard represented by several code sequences (obtained as a result of repeated scanning). The symbol for the contour of the specified object is regarded as having been identified only when there is a complete (or with some admissible tolerances) coincidence of the code combination being compared at least with one of the code combinations of the standard.

Coordinates are recorded, as in the first method, each time the scanning element confronts the contour boundary line; the only difference being that if, after the symbol identification, the coordinates are found not to belong to the given object, they are erased.

The third method

The objective is the same as in the second method. Like the second method, only one symbol from the legend is read into a computer, but in this case the standard is assigned to only one code combination of lines of differing parameters with the indication of the limits of permissible deviations in the parameters of lines for the identified symbol and the standard. The limits of these deviations may be established after analysing the whole legend of a map. A symbol is regarded as identified only when the discrepancies between the codes of the map and the standard lie within admissible limits.

For this method an algorithm and computer program for a fully automatic identification of raster symbols, including the readout of the original map, has been developed and realized in practice. Identification of symbols in accordance with the second method has much in common with the first method described above. However, it has an essential difference which makes its application in solving certain problems more effective. The difference is that, instead of standardizing the whole symbol of the map legend (i.e. all the lines making up a combination), the parameters of each line are taken separately but, in this case, for all the lines used in the map legend. Such standardization makes it possible to get from the map the data on individual characteristics which, taken together, constitute a certain phenomenon. In this case, for the identification to be performed it is sufficient to assign the computer the code of only one line designating the given characteristic. When it becomes necessary to obtain data regarding the area (the coordinates of its boundaries) of the occurrence of the phenomenon as a whole, the operator, using the data of the map legend, inputs the combination of codes for the whole symbol denoting the specific phenomenon. The standard thus formed is used to identify the whole symbol of the phenomenon in question and determine the area of its occurrence.

In conclusion it should be noted that the second method of symbol identification has the advantage of making it possible to read out the data selectively, both for the individual characteristics composing a complex phenomenon and for the whole complex.

4.2.2. The algorithm of identifying the contour symbols by raster line combinations

1. Objective of the algorithm

The objective of the algorithm is to obtain the coordinates of all the boundary points of the contours filled with a specific combination of raster lines. The algorithm consists of two main parts: recognition of raster line combinations; and identification of the coordinates of contour boundary points.

The procedure of tracing the coordinates of the boundary points of a contour is dealt with in section 4.4.

To assure promptitude of data processing the following important condition was set: the data must be processed at the moment of readout, i.e. the machine processing cycle should not be longer than the time needed for the reading of one line. In this case all the data on extraneous contours are ignored.

2. Identification of a raster combination

Restrictions imposed on source information:

1. All contours are separated by a continuous boundary whose colour differs from that of any of the raster lines.
2. There exists a number $a > 0$, such that any distance (defined as the number of points) between the neighbouring raster lines making up a combination is smaller than a, and any distance between adjacent combinations is larger than a.
3. The code of the colour component corresponds to its colour coordinates $(\bar{x}_i, \bar{y}_i, \bar{z}_i)$.
4. The colours and thicknesses of raster lines must be different.

When these conditions are met it is possible to identify raster combinations, as follows. A section of a map with the standards of raster line colours and thicknesses is read out. The set of standards includes a set of all the colours and thicknesses of raster lines actually occurring on the map. The colour of the boundary line is either specified once and for all or represented by an individual standard in the set of standards. For each standard line, from the results of reading scores of lines, are calculated the mean values of colour and thickness components (C_i^k and T_j, $k = 1, 2, 3$) and the maximum deviations (δC_i^k and δT_j). The thickness of a raster line on the map, T, is considered to be coinciding with the standard T_j, if $T_j - T < \delta T_j$. Similarly, colour component, C^k coincides with the standard one, if $C_i^k - C^k < \delta C_i^k$ ($k = 1, 2, 3$). From this immediately follow the requirements for discernibility by colour: for any $i \neq j$ there exists a value of k when segments

$$[C_i^k - \delta C_i^k; \quad C_i^k + \delta C_i^k]$$
$$[C_j^k - \delta C_j^k; \quad C_j^k + \delta C_j^k]$$

do not intersect, and for discernibility by thickness: for any $i \neq j$, when segments

$$[T_i - \delta T_i; \quad T_i + \delta T_i]$$
$$[T_j - \delta T_j; \quad T_j + \delta T_j]$$

do not intersect.

The values of δC_i^k and δT_i (or rather the probability distributions of these values) determine the permissible number of different colours and thicknesses for this technique.

3. Prescribing the raster combinations

Let us assume that we have L possible colours, M thicknesses, and N gradations in a raster combination. It is suggested that the raster combination should be prescribed as follows.

The set of contours, whose boundary points are to be distinguished, is prescribed by several sequences in the following form:

$$i_1, \; j_1, \; k_1$$
$$i_2, \; j_2, \; k_2$$
$$i_m, \; j_m, \; k_m$$

where i, j, k are combinations of ordinal numbers of the colour, thickness, and position, respectively, logically joined by $\cup$, $\cap$,] symbols ($\cup$ = logical 'OR', $\cap$ = logical 'AND',] = logical 'NOT'). For instance, if a set of the standards of three colours and four thicknesses (numbered) has the form shown in Fig. 4.6a, and the assigned sequence is

I	II	III
1, 1, 1	2, 3, 2	2, 2, 3
$i \; j \; k$	$i \; j \; k$	$i \; j \; k$

it means that the contours identified are denoted by a combination, where position 1 is occupied by the line with colour #1 and thickness #1, positions 2 by the line with colour #2 and thickness #3, and position 3 by the line with colour #2 and thickness #2 (Fig. 4.6). A more complex example:

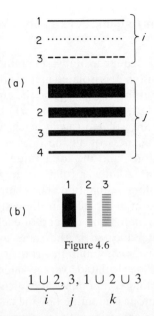

Figure 4.6

$$\underbrace{1 \cup 2}_{i}, \underbrace{3}_{j}, \underbrace{1 \cup 2 \cup 3}_{k}$$

means that the identified contours are those for which at least in one of the three positions there is a line with thickness #3 and colour #2 or #3.

The combination to be identified is introduced manually from the operator's console prior to the program initiation.

4. Operation of the program

During each processing cycle the computer memory contains: ith line—being processed, $(i - 1)$th line—auxiliary, $(i + 1)$th line is introduced while the ith line is being processed. At the end of processing, the ith line is shifted to the place previously occupied by the $(i - 1)$th line and becomes the auxiliary line, the $(i + 1)$th line is transferred to the place of the ith line and becomes the processed line.

In each processed line are singled out series of points with codes differing from the code of the boundary (we assume that the outer boundaries of the map have the code coinciding with that of the boundary between the contours). We assign these series the numbers of $1, 2, 3, \ldots$, etc., going from left to right (in the direction of scanning) and shall subsequently use these numbers for the identification of contours.

We thus have a data set the number of whose element coincides with the number of the singled-out series, and the element itself represents the address of the beginning of the data area for each contour until it has been identified. Moreover, each element is given an index register with an index $r_1 = 0$ if the contour has not been identified and with $r_1 = 1$ if the contour has been identified, and a register with an index $r_2 = 0$ if the contour that has been identified does not enter into the set of those selected, and $r_2 = 1$ if it enters this set. Thus, for each isolated series an element of the data set (we shall designate it as the controlling element) is found by its number. What takes place after that is as follows:

1. If $r_1 = 0$, the maximum sequence of raster combinations (proceeding from requirement 2 in paragraph 2) is isolated in the series, beginning and ending with a gap (i.e. having at the beginning and the end of the series of points a background whose length is not less than a). In each combination the individual raster lines are isolated (again using requirement 2 from paragraph 2). The total thickness of all the lines is computed for the given sequence of combinations, and the total value of codes is computed for every colour component. All the calculations are performed separately for each position of the raster combination. The computed values are added to the previously found sum and placed in the assigned memory area. The coordinates for the beginning and the end of each selected series are also placed there. A counter located in the same memory area records the number of identified raster combinations. If the counter value has reached the preassigned value the calculated average value of the thickness and the code of each of the colour components are calculated. The r_1 value is set equal to 1, and the contour is identified by the characteristics presented in paragraph 2. If the code coincides with the preassigned one (see paragraph 3), r_2 is set equal to 1, if not, it is set equal to 0. If $r_2 = 1$ the already accumulated coordinates for the beginning and the end of the isolated series are transferred into the buffer memory area and then, as the latter is being filled, to the tape.

2. If $r_1 = 1$, the r_2 value is checked. If $r_2 = 0$, the next series is processed. If $r_2 = 1$, the initial and end coordinates of the processed series are transferred into the data area where the results are stored.

4.2.3. Reading raster maps with a specialized device

Specialized devices can identify symbols both by individual raster parameters and by their various combinations. They are designed to raise the identification reliability and provide an opportunity for the computer to solve other problems more effectively.

A specialized reading device is proposed here to identify and select the coordinates of the boundaries of assigned contours, which are then ordered (traced) by the computer.

Figure 4.7 shows the block diagram of this reading device. The logical units shown on the diagram make it possible to identify combinations of lines having not more than three thicknesses and three colours of different brightness. Let us now examine the operation of this device. A raster map is placed in the scanner, S, in such a way that the scanning element should move perpendicularly to the raster lines. The reading head, consisting in this case of one photosensitive element, converts light signals into voltage. Through one of the channels the signals arrive at two amplifier–shaper stages (ASS). The output of

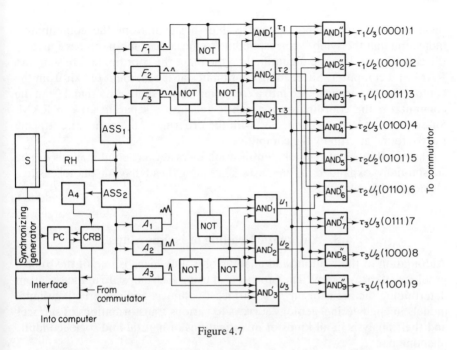

Figure 4.7

the first one, ASS_1, is connected to three filters, F_1, F_2, F_3, for pulse widths τ_1, τ_2, τ_3 (corresponding to three different line thicknesses, T_1, T_2, T_3). The filters, together with logical 'NOT' and 'AND' gates form a unit selecting the pulses by their width (duration). The outputs of logical elements I_1, I_2, I_3 are connected to nine logical 'AND' gates. The output of the second ASS_2 is connected to three amplitude discriminators, A_1, A_2, A_3, that, together with the logical 'NOT' and 'AND' blocks, produce signals proportional to the pulse amplitudes at the output. Elements I_1', I_2', I_3' are also connected to nine logical 'AND' gates.

As a result, at the outputs of logical elements I_1''–I_9'' the following identification signals are formed:

I_1''—signal $\tau_1 U_3$ for the combination of minimum thickness T_1 with brightness B_3,

I_2''—signal $\tau_1 U_2$ for the combination of minimum thickness T_1 with brightness B_2,

I_3''—signal $\tau_1 U_1$, and so on.

All the nine channels are connected to a commutator which sets a certain code combination answering to the combination of signals coming from these channels.

Coordinates of the boundary lines are recorded as follows. When the scanning element confronts the boundary line of a contour (usually having a black colour), at the output of amplitude discriminator A_4 there appears a signal that goes into the coordinates registration block, CRB. This block inquires for the values of coordinates from the pulse counter, PC, and enters them into registers. Upon the arrival of a signal from the commutator, indicating that the symbol corresponding to the code combination assigned by the commutator has been identified on the map, the coordinates are sent into RAM of a computer. Succeeding coordinates at the moment of exit from the contour are also transmitted into a computer. When there is no signal from the commutator the coordinates are transferred into a separate area in RAM. Subsequent operations connected with the ordering of the points, i.e. tracing the contours, are run by the computer.

The possibilities of symbol identification by raster line combinations can be substantially expanded if one uses a reading head having photosensitive elements with colour filters.

4.3. Automatic Reading of Isoline Maps

Automation of the readout and analysis of isoline maps is one of the topical problems of cartography. The prospects of using such maps are immense, from determining the areas of physical surfaces, volumes, and constructing digital models in engineering–geology surveys to various transformations of surfaces and their analysis in all kinds of investigations of natural and socioeconomic phenomena.

This section deals mainly with one aspect of this problem: computer-assisted reading of isolines with the aim of constructing digital models.

4.3.1 Reading normalized isoline maps with a computer

The data in the form of coordinates x, y, z can be read out from normalized isoline maps at any trajectory, recti- or curvilinear, of the scanning head. A map is usually scanned by successive lines, and individual profiles for every pass are plotted to develop a digital model [111, 120]. Various reading devices can be used for this purpose, operating on-line with a computer or in an off-line mode. For the first coding method, described in section 3.2.1, a multipurpose single-channel input device can be used, quantizing the image by its brightness. For the second method a three-channel device is required.

When conventional single- and three-channel input devices are used for polychromatic images, the data obtained have to be processed by a computer. A simple algorithm for the construction of a digital model is examined below.

We assume that the data are read out from the map and inserted into a computer as a regular network of points with the image quantized into n brightness levels. Data processing consists of three basic operations:

1. identifying the isolines by brightness;
2. determining the values of the z-coordinate for the points where the scanning line intersects the isolines;
3. constructing a regular network of points with the values of z.

Identification of isolines by the combination of brightness codes (designated above as signals 1, 2, 3) is performed as follows. At all the points where scanning lines intersect the isolines there will be a continuous series of points u_i whose brightness code values exceed those for the codes of the background points. The brightness code u_k of isoline, i.e. signals 1, 2, or 3, are found as the mean value of the codes for all the adjacent points, u_i, belonging to the given isoline. The obtained values of $\bar{u}_k$ signals are then compared with three standard (reference) signals: the regions of code values $\bar{u}_0 \div \bar{u}_1$, $\bar{u}_1 \div \bar{u}_2$, and $\bar{u}_2 \div \bar{u}_3$, assigned to the three levels of brightness or optical density.

The code of isoline brightness can be determined more accurately by another method. It consists in an exhaustive search for the point with the maximum brightness code value in a group of adjacent points belonging to the isoline in the given scanning line. The value for this point is regarded as corresponding to the brightness code of the isoline. This code is then identified, as in the previous case, by comparison with the reference regions.

For reliable identification of isolines by three standard signals it is necessary that the scanning element should hit the isolines at least twice over their thickness, and that there should be at least a two-fold difference between the brightness levels of isolines and the background.

The procedure of isoline identification described above is only suitable for the first coding method. In the second method similar functions are performed by a three-channel reading unit. All the subsequent computer data processing procedures are similar for both methods. The absolute values of the third spatial coordinate of isolines is determined starting from the left or the right inner edges of the map, depending on the scanning direction. The value of a point at the edge of the map is equal to the absolute value of the z_0 coordinate of the initial point of the first isoline coinciding with the edge. The z_j coordinates of the initial point of subsequent isolines is equal to $z_0 + \delta \Delta z$, where δ is a coefficient whose values, depending on the form of signal transfer, can be equal to $+1$ or -1, or 0 (in the presence of half-interval contours, $\delta = \pm 0.5$). The change of the sign of the coefficient in passing from one isoline to another along one scanning line, as well as in passing from one line to another at the initial scanning points, is the same. Thus, in transitions

$$
\left.\begin{array}{lll}
1 \to 2; & 2 \to 3; & 3 \to 1; \quad \delta = +1 \\
3 \to 2; & 2 \to 1; & 1 \to 3; \quad \delta = -1 \\
1 \to 1; & 2 \to 2; & 3 \to 3; \quad \delta = 0
\end{array}\right\} \tag{4.7}
$$

The value of the initial point for the nth isoline is equal to

$$
z_j = z_0 = \sum_{k=1}^{n} \delta_k \Delta z \tag{4.8}
$$

The value of the ith point of intersection between the isoline and the scanning line is obtained from formula

$$
z_{ij} = z_j + \sum_{k=1}^{i} \delta_k \Delta z \tag{4.9}
$$

Interpolation of z values needed for the construction of a regular digital model is performed after all of them have been determined at all the intersection points. In linear interpolation along the jth line the $z_{k,j}$ coordinate in node k, j with the $Y_{k,j}$ coordinate at the regular grid is determined from formula

$$z_{kj} = z_{ij} + (z_{i+1,j} - z_{i,j+1}) \frac{y_{k,j} - y_{i,j}}{y_{i+1,j} - y_{i,j}} \qquad (4.10)$$

where $z_{i,j}$, $y_{i,j}$ and $z_{i+1,j}$, $y_{i+1,j}$ are the corresponding values of the coordinates of points (i, j) and $(i + 1, j)$ at which the jth scanning line intersects the adjacent isolines with the nodal point (k, j) located in between. When plotting the grid it is unnecessary to assign the density of the nodes equal to the density of the scanning lines. Depending on the problem being solved, the distance between the nodal points of the grid can be taken arbitrarily, but it should be divisible by the line advance of the scanner.

To obtain a more accurate digital model it is necessary to apply other interpolation methods (see section 4.3.4).

The methods examined above have the following advantages and drawbacks. These methods, especially the one with a latent optical code, make it possible to compose normalized isoline maps not differing in appearance from traditional maps. The isoline values being identified in a unidimensional space, a high promptitude is achieved with a rather simple program of data processing on a computer or with the help of a specialized unit.

The only drawback of these methods is that they somewhat complicate the technology of map publication. Instead of one plate usually needed for the printing of isolines, three such plates are required in this case. Printing of the other elements of maps is the same. The use of fluorescent paints makes the cost of maps somewhat higher. A certain complication of map production is, however, compensated by the economic effect achieved with complete automation of the process of developing a digital model based on normalized isoline maps.

It should be noted that there exist various methods of converting an isoline map into a digital model with the help of specialized devices and a computer, based on the marking of isolines, but all of them have serious deficiencies. The method of marking isolines, as applied to the operative test model 'Map', described in reference 165, consists in drawing the isolines as two closely adjacent lines of different colour (red and black). Not only is the traditional appearance of a map disturbed in this case, and the visual clarity decreased but, what is more important, the accuracy of image representation is diminished. The method of drawing adjacent isolines with two paints of different colours also has two serious defects: (a) it decreases the visual clarity of images; (b) it requires more complex programs for computer processing because the image has to be analysed in a two-dimensional space. Moreover, the above methods have another common defect: when they are used no images except the isolines themselves are allowed on the map.

4.3.2. Reading normalized isoline maps with a specialized device

Let us examine the principle of reading isolines with a latent optical code (following the method presented in section 3.2.1) using a specialized device [58, 120].

Figure 4.8a shows the block diagram of the reading unit. The general view is given in Fig. 4.9. The unit contains a scanning block I made in the form of a drum. Scanning is performed with the help of two step-motors (2, 3) by means of the reciprocating motion of the drum with a map, i.e. with the drum rocking along the y axis and the reading head being displaced with respect to the drum along the x axis. When the whole area of the map is being scanned the reading head is displaced with the drum placed in the extreme positions. In other words the map is scanned in such a way that the scanning spot in its reciprocating motion does not go beyond the field of the map. Such scanning can be called continuous (Fig. 4.8b).

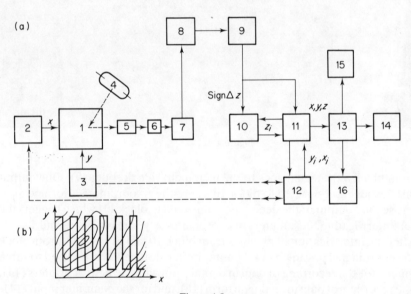

Figure 4.8

When the scanning spot confronts an isoline, the latent optical code is decoded with the help of an ultraviolet radiation source and a quartz condenser (4). The generated luminous radiation (luminescence), passing through an objective lens (5) and a dividing unit (diffraction grating) (6) enters the receiver (7) consisting of five photomultiplier tubes.

Combinations of isolines with different spectral characteristics of paint luminescence produce different combinations of signals at the reading head output, which then enter the signal-forming block 8.

Figure 4.9

From block 8 the signal goes to the increment sign decoder (9). One output of this decoder is connected to block 10 where the z_i coordinate is formed.

The second output of the decoder is connected to block 11, which triggers the circuit interrogating the counters of coordinates x, y, z in blocks 10 and 12.

After the interrogation of z_i values from block 10 and x_i, y_i values from block 12 the coordinates passing block 11 enter buffer storage 13, connected to three channels: for the recording of data in digital form with the help of printers (15), for the recording of data in analogue form (16), and for the computer input (14).

One of the basic blocks of the device is the decoder (9). The unit can decode three-, four-, and five-spectral codes. From the point of view of the maximum informational reliability, the four-spectral sequence of isolines whose paint compositions contain luminophors with three different spectral characteristics should be regarded as the optimal code combination. Such, for example, are achromatic (grey) compositions with a glow (at the exposure to ultraviolet radiation) in the blue–violet, yellow, and red spectral regions. The fourth spectral characteristic is obtained by mixing all three luminophors together. This kind of code combination assures the decoding of the increment sign, Δz_i, for all possible sequences of four-spectral isolines.

A possibility is also provided to form a double increment, when the next-in-turn isoline is skipped, either by chance or intentionally, at places where the isolines are highly congested.

4.3.3. The methods of normalizing traditional isoline maps for automatic construction of digital models

When developing digital models based on topographic, geophysical, and other traditional maps it is recommended, from the point of view of an optimal division of functions between man and computer, that the following principal operations be performed:

1. preparation of source data for the input into a computer, i.e. the normalization of isoline map;
2. processing of data introduced into a computer: identification and description of objects with their normalization taken into account (noise removal, determination of the elevations, and relative positions of isolines, etc.);
3. construction of a digital model using a specific interpolation technique with the help of appropriate software.

This section examines the first principal operation, i.e. questions related to the normalization of traditional isoline maps [131]. Analysis of the possible ways of automating the construction of digital models proceeding from the currently available (traditional) isoline maps shows that at present two methods of solving this problem are being developed:

1. Recording the coordinates of isolines and their values on a magnetic storage medium (tape or disk) by tracing them manually using a digitizer, and then processing these data on a computer.
2. Preparing the original map by means of special techniques of isoline marking with the aim of simplifying the automatic identification and raising its reliability, followed by scanning the map, data input, and computer processing.

The first method is very laborious and painstaking. It is very time-consuming and, at the same time, does not guarantee a sufficiently accurate readout, since the isolines are traced manually. According to reference 110 it takes about 200 man-hours to digitize a medium-mountainous relief on a 1:50,000 map. The second method has certain advantages. It is, however, also very laborious, since the presently used marking techniques involve complete outlining of each isoline in a certain colour in accordance with specially formulated rules. This explains the relatively low effectiveness of this method [66].

Special experiments have made it possible to arrive at a conclusion that the following two approaches are the most expedient and promising in solving the problem of constructing a digital model, based on the available isoline maps, with the help of a computer:

1. automatic construction of a digital model after a partial manual preparation of the original isoline map;
2. semi-automatic construction of a digital model without the initial preparation of the original isoline map.

Within the framework of the first approach, four methods are recommended to prepare maps for the automatic reading and processing with the help of scanning devices and a computer:

1. multi-feature marking of extreme points;
2. multi-feature arbitrary marking of isolines;
3. even-numbered marking;
4. marking of bergstrichs.

The second approach is best realized by the following two methods:

1. data processing in the dialogue mode with a graphical display;
2. the use of specialized tracing devices.

We shall start with discussing the last two methods. The first one of them presupposes the initial stage of scanning the original isoline map (a colour-separated negative or positive), followed by tracing the isolines (in the main, this tracing algorithm and the algorithms of tracing in all the methods related to the first approach are similar), output onto the screen of a display with a light pen, visual identification of each isoline using a light pen and a keyboard, and then constructing a digital model.

We see here obvious similarities with the methods related to the first approach: using special rules one can sharply reduce the work with a light pen. Of interest are the suggestions made in reference 34. Preparation of isoline maps by traditional graphical means is, nevertheless, much cheaper and simpler: only a part of the processed map sheet can be shown on the display screen, and the software for the scanning of maps with the use of a display and a simultaneous marking of isolines is rather complicated. An obvious advantage of the method is the possibility of resolving indeterminate situations when tracing complex objects in the dialogue mode. Note that in the methods related to the first approach retouching is used for this purpose. However, in these methods too, if a computer is equipped with a graphical display it is expedient to use it; although, as our experience shows, indeterminate situations during processing are comparatively rare in this case.

The second method consists in using specialized tracing devices of the 'photoeye' type for the input of isolines into a computer. With respect to labour expenditures it is clearly inferior to the methods based on scanning. One should take into account two other important things. The tracing unit operator works under stress and can therefore make mistakes (omission of isolines, erroneous continuation of isolines after a gap, etc.) which requires additional checking, or repeated data input (with a rather complex software for the comparison of similar graphical objects in computer memory), or else automated drawing of isolines to be visually compared later.

We shall concern ourselves below with the methods related to the first approach, as the most promising and distinguished by high efficiency, promptitude of processing, and reliability of isolines identification. All these methods incorporate the procedure of partial normalization (i.e. manual marking) of the original map.

The method of multi-feature marking of extreme points
The method proposed is closely related to the method of complete normalization of isolines which can be used in printing new maps or updating the already existing ones [111]. It will be recalled that the gist of the method is as follows. Isolines are given on a map with paints of such a colour (or brightness) that makes them reliably distinguishable against the background of the map in the course of scanning. By these features the isolines are divided into three classes: 1, 2, 3. Alternation of classes $1 \rightarrow 2 \rightarrow 3 \rightarrow 1$ of the neighbouring isolines indicates an increase of the function, and the inverse alteration indicates its decrease. The principle of three-feature marking is easily extended to cover the case of four, five, and more features. It is obvious that this principle assigns only the relative values of elevation; therefore to determine the absolute values of all isolines it is necessary to specify in advance the elevation of a certain (e.g. the first) isoline on the map. The algorithm of tracing the isolines for the construction of a digital model proceeding from fully normalized maps is treated in reference 143. One of the advantages of such maps is the possibility of rapid linear interpolation along a line in the course of scanning. Construction of the digital model can be assigned to the specialized unit of a scanner. This approach is wholly realized by the F-140 scanning device (see section 4.3.2).

The described scheme and, what is very important, the developed software can be used in processing traditional isoline maps. For this purpose the original map has to be manually prepared beforehand. Specially prepared paints are used to place small signs (markers) of an arbitrary form on the maximum points of isolines (Fig. 4.10). When the map is scanned from top to bottom, the maxima of isolines, i.e. the markers, will be processed first. This will immediately show that the isoline belongs to a certain class. In the course of tracing this feature (sign) will be transferred from the maximum point to all the other points of the isoline.

As compared with complete tracing of isolines (outlining them) the partial marking of the extreme points simplifies the process of map preparation. The paints for marking are specially selected for the specific input device to be used. If the device provides for a multicolour input the negative is marked with transparent chromatic paints; if the input is monochromatic it is marked with grey colours of varying optical density. The marking can be performed on monochromatic print from a negative by photoreprographic or polygraphic means. With a monochromatic input, paints of differing brightness are taken for the marking of such prints; with a multicolour input chromatic paints are used with widely differing spectral characteristics.

Signs can be marked with fluorescent paints which, as shown by experiments,

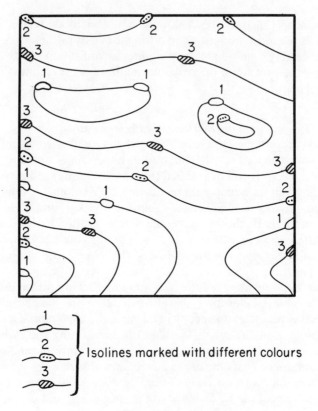

Isolines marked with different colours

Figure 4.10

provide the highest reliability of isoline identification. A negative is marked with fluorescent paints having widely differing spectral characteristics, following any principle of coding. The prepared negative is superimposed on a luminescent base (screen) whose radiation spectrum differs from those of the marking paints. The input of the map into a computer is accomplished by an appropriate scanning device (e.g. F-140).

Multi-feature arbitrary marking of isolines

This method is designed for the complete tracing of all the isolines on a map, when after the tracing the image of every isoline is stored in computer memory. In this case it becomes possible to improve the interpolation method and simplify the procedure of manual marking.

Each isoline is marked only once at an arbitrary point. The idea of coding is the same as in the previous case (Fig. 4.11). If, when the contour of an isoline is being traced, it is interrupted by a marker the identified code is assigned to the whole isoline. For convenience some isolines may be marked more than once; it is only required that the codes of all markers coincide.

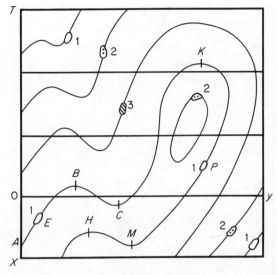

Figure 4.11

With complete tracing there arises the possibility of introducing a convenient method of dealing with some situations difficult for algorithmic presentation: ravines, gullies, bluffs, taluses, steep slopes, and inscriptions encountered on the map are 'retouched', i.e. filled with a paint whose brightness or colour differs from that of the marker paints. In the course of tracing, such retouched areas are regarded as independent objects where the data have not been determined. When the data are processed it is possible to identify the isolines lying at the boundaries of the retouched areas and extend them, following a certain method, into the areas. It should also be noted that there is the possibility of marking not all the isolines but only those whose elevation is divisible by a certain step—e.g., only the thickened isolines. The larger the step, the fewer the marked isolines. Unmarked isolines are excluded from further processing, which allows the source data to be simplified. This technique is convenient for maps with relatively small contour intervals, e.g. for high-mountain relief.

Even-numbered marking

Let H be the contour interval. Only one paint is used, which marks isolines having even-numbered H with all the other isolines remaining unmarked (Fig. 4.12). Additionally assigned are the value of the first isoline and the direction of elevation increase in its vicinity. Retouching can be performed with the same paint as the marking if the areas of markers and retouches are reliably discriminated.

The main advantage of this method is that the operator, when preparing a map for scanning, does not have to change the writing instrument alternately, which was required in the previous methods. On the other hand it entails

Figure 4.12

certain complications in software, since in the previous methods the very fact of the presence of markers on the isoline made these objects readily distinguishable against the overall background of the map.

Marking of bergstrichs

We assume that, as a result of computer processing, all the extreme limiting points of a function at the inner edge of the map have been found.

If the types (minimum, maximum) of these points, the elevation of the preassigned (initial) isoline on the map and the bergstrichs of closed isolines are also known, the elevations of the other isolines are then determined unambiguously.

An obvious solution of the problem thus formulated lies in normalization. Bergstrichs on closed isolines are 'marked'. This operation is promptly and reliably performed even by an operator who cannot read isolines. Let us now turn to the extreme boundary points. Generally speaking, the conventional data on bergstrichs are no longer sufficient here. Indeed, to determine the elevation of a specific isoline lying at the edge of a map one often has to make use of absolute elevations, bergstrichs, the outlines, and relative positions of isolines. The extreme points along the map edge should therefore be explicitly shown. It is, however, unnecessary to assign (code) their type, since they alternate on the map edge: a maximum is followed by a minimum, and so on. A bergstrich marked on any open isoline will determine the type of a certain boundary extremum and, consequently, of all the boundary extremums. Thus, markers along the map edge are placed in between equivalent isolines. Another

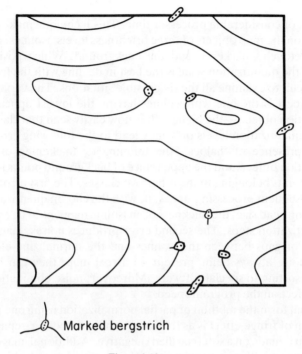

Marked bergstrich

Figure 4.13

simplification is permissible: it is unnecessary to place such a marking dot at the map edge between two ends of one isoline. The presence of an extremum here is readily identified by means of a computer program (Fig. 4.13).

Depending on the normalization method and other preset conditions (continuity or smoothness of a function, real-time processing, etc.), different interpolation techniques can be used in data processing (see section 4.3.4).

Elimination of the fringe effect
In our case the fringe effect is understood as the appearance, along the boundary of an image belonging to one optical class, of points (or 'haloes') whose optical characteristics belong to another class [14]. This effect is associated both with the weak pressure of the writing instrument at the edges of objects (e.g. in marking), and with the scanning head only partially confronting the image boundary. Because of the fringe effect certain difficulties arise: it becomes impossible to refer these indeterminate points to one of the two classes (or to the background) simply by choosing the quantization levels.

If the normalization method employs only one point, a technique of map preparation (normalization) can be suggested, with which the fringe effect has practically no influence on the processing. The use of this technique is highly effective as it obviates the necessity of solving this problem by means of a computer program, which would require a large area in RAM and an intricate algorithm, or considerable processor time. This technique makes it possible to single out on the map objects of three brightness levels: isolines, normalization objects (retouch, markers), and the background. Which position of these classes on the quantization scale is the best in dealing with the fringe effect? It is not difficult to examine all the six possible situations. The variant that proves to be the best is the one with isolines having the lowest optical density and normalization objects—the highest. It is now easily seen that classifying all the points by quantization levels only may lead to the following errors associated with the presence of 'haloes': the thinning or thickening of isolines and normalization objects and the appearance of background points in between the adjacent points belonging to the other two classes. The first error has no effect on the subsequent processing stages, if, with the chosen quantization level and the scanning head size, the thickness of an isoline happens to be equal to at least two discretization steps. The second error produces noises, manifested either by 'holes' or gaps between the isolines and the normalizing elements. Such types of noise always being present on a real map, they can be eliminated without resorting to special software. Moreover, noise elimination has practically no effect on the program speed.

In its final form the method of partial normalization using one paint with the elimination of fringe effect is as follows. A copy of a colour-separated original isoline map is made on a soft (clarified) negative. Additional images are applied on the negative with one paint in accordance with a specified rule. The paint must be optically more dense than background (e.g. ordinary retouching can be used).

In conclusion it should be noted that the described methods of automated construction of a digital model, preceded by a partial manual preparation of the original isoline map, based on marking the isolines with one or two short strokes, have significant advantages over the presently known methods:

1. a higher accuracy of digitizing the isolines is assured, as these are traced by a computer (in contrast to manual outlining accompanied by numerous errors);
2. the time needed for preparation and digitizing is substantially reduced, since the procedure of complete manual tracing of isolines is excluded;
3. conditions are provided for expanding the scope of work by using a large number of operators to prepare (normalize) the initial isoline maps for automatic processing, which cannot be done within the framework of the currently applied digitizing methods, where each operator must have his own digitizer;
4. a higher economic efficiency is achieved.

4.3.4. The methods of computer-assisted interpolation of scanned isolines

When choosing the method for the interpolation of scanned isolines, one has to take into consideration the following factors:

1. the complexity of software for the implementation of the method on a computer;
2. the required computer resources (memory, processor time);
3. the method of normalizing the initial data;
4. the type of the digital model being constructed.

Let us examine the following classification of interpolation methods: the methods of unidimensional interpolation, the net methods, and the methods of two-dimensional interpolation. Within a certain class all the methods impose identical requirements upon the structure of input data, i.e. the tracing algorithm. The methods of interpolation are considered below in the context of both the type of tracing and the type of normalization.

4.3.4.1. *The methods of unidimensional interpolation*

Included in this class are the methods whose realization requires only the data on the coordinates of isoline intersections by the scanning line and the elevations of corresponding isolines. On the strength of these data, the relief function value is restored along the given lines. These data for fully normalized maps can be prepared by the specialized equipment alone [78]. Similar data for other normalization methods are obtained with the help of special tracing programmes. For instance, in processing the isoline maps with marked extreme points a comparatively simple tracing technique is used, following one line after another. An identified marker is attributed to the whole isoline. Such local tracing needs no complex data base and can in principle be realized by means of the same software and equipment.

The methods of unidimensional interpolation are thus developed and applied when, for some reason or other, simple tracing is used (e.g. when computer resources are limited). A high quality of initial material is naturally required in this case: absence of noise, of discontinuities, etc. If the noises existing on a map are to be treated in a computer program, the use of more complicated and more suitable interpolation methods will be warranted.

It is not difficult to see that the more convenient the normalization, the more complicated the tracing it presupposes. An intermediate position is occupied by the method of multi-feature arbitrary marking. On the one hand, it was developed for general isoline maps, without any specific requirements to the quality of image, and therefore presupposes complicated tracing. On the other hand, if the quality of initial material is good (in particular, if no retouching has been used) the method can also be applied with simple tracing. Its hardware implementation is also possible, though it will be more complicated than in the case of marking the extreme points. As an illustration, a brief description of this type of tracing is given below.

The idea of the method is illustrated by Fig. 4.11. Here a portion of an isoline map is normalized by means of three-feature marking (a feature is convention-ally designated by a digit). The tracing, as well as the input of the map into a computer, is performed linewise from top to bottom. When portions of isolines are processed in the vicinity of points B, K, and H, the global type of isolines, specified by markers E and P, is unknown. It is therefore assumed that at point P a marker of the T_1 type, at point K of the T_2 type, and at point H of the T_3 type are placed. These markers are 'spread out' onto all the points from top to bottom, so that ABC will have the T_1 type, and so on. At point E, when the true marker is identified, we shall obtain the $T_1 = 1$ equality, and at the points where portions C and M are spliced: $T_1 = T_2$ and $T_3 = T_2$. Consequently, when the tracing is completed, all conventionally set markers will acquire specific values. Now suppose that we have to restore the relief along the OU straight line. While this line is being traced the coordinates of its intersections with the isolines and the types of markers (conventional) for these isolines are memorized. For OU we shall have: T_1, T_1, T_2, T_2,... Similar data are memorized for the left edge of the map, TX. When interpolating, we retrieve from the memory the data for the TOU route, the true types of markers are substituted for conventional ones, elevations are then recalculated in the usual way, after which the formulae of unidimensional interpolation are used.

A more formal description of this algorithm can be proposed, making it suitable for hardware implementation: in the memory is allocated Area T, sufficiently large to store a line of the initial map whose every point is an integer. In the course of scanning, each subsequent point acquires one of several values: B when the scanning head confronts a point of the background, I a point of an isoline, M a point of a marker (in actual fact, M_1, M_2, etc., according to the number of features marked). Line T will be regarded as having been processed if every point on it is either the background or the conventional number, N, of an isoline, which in tracing is spaced downwards along the isoline (as number 1 for the ABC portion of the isoline in Fig. 4.11). When a new line is being scanned the previous line T is transformed in the memory into the new one and processed in three stages.

The first stage—replacement: conventional numbers of isolines are transferred to the new line and the encountered marker points are assigned to the corresponding conventional numbers. Let pair $\{T(i), P\}$ designate the ith point of the previous line T and the ith point of the line being read out. Since point $T(i)$ can have one of two values, B or N, and point P can have one of three values, I, M, B, such a pair describes six possible situations. Response to each of them at this stage is as follows. The (N, I) pair is not processed, i.e. point I on the isoline acquires the number of N. The (N, M) pair indicates that a marker point M has been encountered on the isoline with conventional number N. Every conventional number has in the memory the corresponding counters for each marker type, fixing the number of points of this type present on the isoline. After the tracing has been completed by these counter-values the true type of the isoline marker replacing its conventional number is determined. In the course of processing the above-mentioned pair the counter of marker M, belonging to the conventional number N, is raised by 1. For the four remaining pairs the processing is described by assigning $T(i) = P$.

The second stage—processing from left to right all the pairs of adjacent points T: $T(i)$, $T(i + 1)$. Point T assuming here one of four possible values (I, M, B, N) results in 16 different states, of which only three require processing. The (N, I) pair: two points of the isoline are located next to one another, one of these points has the number of N, while the other one has not yet been assigned its number. The processing is evident: $T(i + 1) = N$. As in the first stage, the processing of (N, M) consists in raising the value of the M counter for number N. Finally, the case of different numbers involves the already mentioned splicing procedure (point C in Fig. 4.11), when equivalence of numbers N_1 and N_2 is memorized. In subsequent determination of the real number of the marker for a given isoline the counters N_1 and N_2 are summed up.

The third stage—processing from right to left the pairs of neighbouring points $T(i - 1)$, $T(i)$. As in the previous case the processing of only non-trivial pairs is described. The (I, N) pair: $T(i - 1) = N$; the (M, N) pair: M is written into the counter of the conventional number N. As compared with the previous situations, the processing of the (I, B) pair is different, in that the isoline in question has not yet been assigned its conventional number (e.g. the point of maximum K in Fig. 4.11). A new number N is generated here, and $T(i - 1) = N$ is assigned. The (M, B) pair is processed in a similar way, but another operation is added: raising the M index in the N number.

The way to obtain the data necessary for unidimensional interpolation has thus been described for the three above-mentioned normalization methods. The simplest interpolation of this class is linear. We shall not dwell on it in detail. It should be noted, though, that for hardware implementation this method, because of its simplicity, is the most preferable. Another method, usually applied in similar situations, is spline interpolation.

Let axis X be the straight line of interpolation (Fig. 4.14), points $x_1, x_2, \ldots,$ x_k be the points of intersection with isolines whose elevations are $z_1, z_2, \ldots, z_k$. From the theory of splines it follows that in this situation there exists a spline of the C^2 class of smoothness coinciding in each (x_i, x_{i+1}) interval with a certain polynomial of the third power. To determine this global spline unambiguously it is necessary to impose two additional requirements on it. With the situation

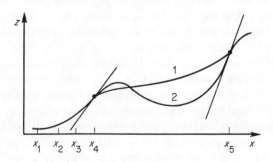

Figure 4.14

in question, splines of such type are inapplicable for two reasons. Firstly, it is impossible here to assign naturally two additional conditions (say, terminal conditions), and the arbitrary assignment of these conditions will lead to a mismatch in the results along the adjacent scanning lines. Secondly, such splines do not convey some of the characteristic features of the approximated functions. Figure 4.14, for instance, shows the relief as represented by curve 1, and the corresponding spline by curve 2, i.e. here monotony is disturbed, as a result of which it becomes difficult to use the obtained digital model (e.g. for the plotting of isolines proceeding from the data of the model, the plotting of equi-illuminance curves, etc.). The use of local splines [4] proves to be more convenient, when first the derivatives are found at points x_i from numerical differentiation formulae, and then, based on these derivatives, a polynomial is constructed on each segment. In [4], for example, a method is described which, applied to the $[x_4, x_5]$ segment (Fig. 4.14), is as follows. The $z'(x_4)$ derivative is sought by five points:

$$z'(x_4) = \frac{1}{p + q} \left(\frac{z_4 - z_3}{x_4 - x_3} p + \frac{z_5 - z_4}{x_5 - x_4} q \right) \qquad (4.11)$$

where

$$p = \left| \frac{z_6 - z_5}{x_6 - x_5} - \frac{z_5 - z_4}{x_5 - x_4} \right|; \quad q = \left| \frac{z_4 - z_3}{x_4 - x_3} - \frac{z_3 - z_2}{x_3 - x_2} \right|$$

a similar formula is used to calculate $z'(x_5)$. The values (numbers) of z_4, z_5, $z'(x_4)$, and $z'(x_5)$ unambiguously define a third-power polynomial, on the $[x_4, x_5]$ segment. It is spliced with the polynomials on the adjacent segments by the first derivative. However, even this improved method has the above-mentioned drawbacks in the case of relief interpolation. The method of constructing a local spline of another type, free of these drawbacks, is examined below.

Let us consider the $[x_4, x_5]$ segment in Fig. 4.14. If $z_5 > z_4$ the function reconstructing the relief has to rise monotonically. In this example the derivatives are selected in such a way that the given function is bound to have an inflection point. From this follows the second requirement, necessary in relief interpolation: the monotonous function constructed must have not more than one inflection point. Let us suppose that there exists a family of monotonously increasing $\psi_{\alpha,\beta}(x)$ functions having not more than one inflection point, assigned on the segment of $[0, 1]$ and assuming the value of $[0, 1]$ with $\psi_{\alpha,\beta}(0) = 0$, $\psi_{\alpha,\beta}(1) = 1$; $\psi'_{\alpha,\beta}(0) = \alpha$; $\psi'_{\alpha,\beta}(1) = \beta$. Then the function, necessary for the local interpolation by four numbers, z_4, z_5, $z'(x_4)$, $z'(x_5)$, is sought in the form of $z = P\psi_{\alpha,\beta}(Q_x + R) + S$. Every $\psi_{\alpha,\beta}$ function is a spline with several inter-mediate nodes on $[0, 1]$ of the power 2, having a continuous first derivative with respect both to the variable and the α and β parameters. Outlined below is the construction of this family, essentially for two-dimensional interpolation.

It is not difficult to see that for relief interpolation the necessary $\psi_{\alpha,\beta}$ family is confined to the case of $\alpha \geq 0$, $\beta \geq 0$. Construction of $\psi_{\alpha,\beta}$ begins with constructing its $S_{\alpha,\beta}$ derivate sought in the form of a continuous piecewise linear

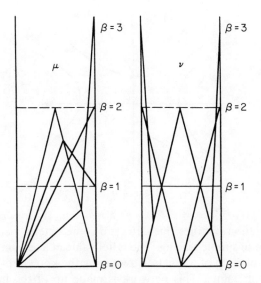

Figure 4.15

function. The $\psi_{\alpha,\beta}$ function obtained by integrating $S_{\alpha,\beta}$ will thus prove to be a continuously differentiable spline of the power 2. The conditions imposed on $\psi_{\alpha,\beta}$ are reduced for its $S_{\alpha,\beta}$ derivative to the form of $S_{\alpha,\beta}(\theta) = \alpha$, $S_{\alpha,\beta}(1) = \beta$,

$$\int_0^1 S_{\alpha,\beta}(x)\,dx = 1$$

and the requirement of having not more than one extremal point at $[0, 1]$. The first to be constructed are the $S_{\alpha,\beta}$ functions for the boundary values of parameters: $\mu_\beta = S_{0,\beta}$, $\nu_\beta = S_{\beta,\beta}$, $\chi_\beta(x) = S_{\beta,0}(x) = \mu_\beta(1 - x)$. The graphs of these functions, clearly indicative of their formal construction, are shown (for several values of the parameter) in Fig. 4.15. The function is finally determined as follows:

for $\beta \geqslant \alpha > 0$: $S_{\alpha,\beta}(x) = \dfrac{\alpha}{\beta}\nu_\beta(x) + \left(1 - \dfrac{\alpha}{\beta}\right)\mu_\beta(x)$

$$\tag{4.12}$$

for $\alpha \geqslant \beta > 0$: $S_{\alpha,\beta}(x) = \dfrac{\beta}{\alpha}\nu_\alpha(x) + \left(1 - \dfrac{\beta}{\alpha}\right)\chi_\alpha(x)$

$$S_{0,0} = \nu_0$$

The indeterminacy existing in the vicinity of the null parameters is eliminated if we set $S_{\alpha,\beta}(x) = \nu_0(x)$ in this vicinity.

4.3.4.2. The methods of network interpolation

The methods of linear interpolation discussed above have an obvious drawback: relief variations are accounted for only in the direction of scanning.

138

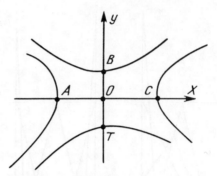

Figure 4.16

Network interpolation rectifies this defect to some extent, permitting at the same time the use of simple tracing. The relief value at a certain point is found in this case with the help of a function averaging the known relief values on a certain network of points. This network is made up of the intersections of isolines with the scanning lines and 'columns'. Only some simple variants, constituting a combination of unidimensional interpolation methods, are used as the average function. It is in principle possible to construct directly from a given network a certain function of two variables (e.g. a local bicubic spline). However, it becomes necessary in this case to perform unnecessarily cumbersome calculations. It seems more effective to improve the quality of relief reconstruction by choosing for every point its own small interpolation network, determined by the shape of the neighbouring isolines. This approach is realized in the methods of two-dimensional interpolation.

The simplest method of this class is bilinear interpolation. Let us designate (Fig. 4.16) the distances from the centre of interpolation O to the nearest isolines along the x and y axes as α_i, and the elevations of these isolines as z_i ($i = A, B, C, T$). The interpolation is then specified by the formula:

$$z(0) = \frac{1}{\alpha_B \alpha_T + \alpha_A \alpha_T} \left(\alpha_B \alpha_T \frac{\alpha_A z_C + \alpha_C z_A}{\alpha_A + \alpha_C} + \alpha_A \alpha_C \frac{\alpha_B z_T + \alpha_T z_B}{\alpha_B + \alpha_T} \right) \quad (4.13)$$

Another variant of the weighted average has the form of

$$z(0) = \left(\sum \alpha_i^{-1} \right)^{-1} \sum z_i \alpha_i^{-1}$$

Splines can also be used for network interpolation. For instance, let it be necessary to calculate the relief value at point (x, y). For this purpose we shall consider straight lines $y = k\delta$, where $k = 0, \pm 1, \pm 2$, and δ is a certain small step. Let us now interpolate along these lines with local splines S_k (the points forming the interpolation network for this method are shown in Fig. 4-17). The value of

$$H = \sum_{k=-2}^{2} p_k S_k(x)$$

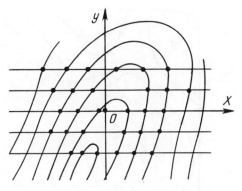

Figure 4.17

for a certain system of weighting coefficients expressing 'the degree of confidence' in the result ($p_{-2} < p_{-1} < p_0 > p_1 > p_2$), determines the relief value at point ($\bar{x}$, $\bar{y}$). Another procedure consists in calculating the K quantity analogous to the H quantity along the y axis, and $z(\bar{x}, \bar{y})$ being set equal to $(H + K)/2$.

4.3.4.3. The methods of two-dimensional interpolation

These methods are based on the isolines being memorized in the course of tracing. The tracing algorithm itself, and the corresponding structure of data, will not be discussed here because of their relative complexity. We shall only note that in implementing these methods it is expedient to use simplified types of normalization.

The first method is based on a technique resorted to by cartographers for visual interpolation. Let O be the point on the map where we are seeking for the value of the function. This point lies within a certain area limited by isolines. As a rule there are two such isolines. In the general case it can be said that either all the isolines limiting the area have the same elevation, or there exist two elevation values. In the second case the interpolation method will be as follows. The shortest distance, $K1$, to the isolines with elevation $H1$ is found. For $H2$ elevation a similar distance will be designated as $K2$. The elevation at point O is then found from the formula of linear interpolation along curvilinear route POT (Fig. 4.18).

$$H(0) = \frac{H1 \times K2 + H2 \times K1}{K1 + K2} \tag{4.14}$$

If point O lies in the area limited only by one isoline and the map edge (area C Fig. 4.18), no plausible conclusion can be drawn on the nature of the behaviour of the function in it. As a first approximation it is recommended that the function here be approximated by a plane. In the general case it is recommended to disregard the boundary points.

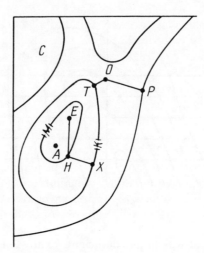

Figure 4.18

Let us now consider the case when the points lie between the extreme isolines. Without going into the details of realization we shall only formulate the general idea. We first look for a point that will be regarded as the extreme point. As a first approximation we can consider point A to represent such a point (or one of such points) within isoline M (Fig. 4-18) lying at the maximum distance from M. A more plausible result will obviously be obtained if subsequent isoline K is also taken into account. In this case a new definition of the distance to M is introduced. According to this definition the distance from point E to point H on isoline M depends upon the distance between H and the next isoline K, and is equal to EH/HK (Fig. 4.18). For the example in question, point E will be the new centre.

It only remains for point E to be assigned a definite elevation value. We shall assume that in the vicinity of point E the relief can be comparatively well approximated by elliptic paraboloid:

$$\frac{x^2}{p} + \frac{y^2}{q} = 2z \tag{4.15}$$

The area of the ellipse obtained by intersecting the paraboloid with the plane, $z = H$, is equal to $2\pi H\sqrt{pq}$. The ratio between the areas of sections is therefore equal to the ratio between their elevations. Applying this fact to Fig. 4.18 we shall obtain: the ratio of areas S_M and S_K, limited by isolines M and K, is correspondingly equal to the ratio of the elevations of point E above M and K. Let h be the contour interval and t be the elevation of E above M. Then:

$$\frac{S_M}{S_K} = \frac{t}{t + h} \qquad t = \frac{hS_M}{S_K - S_M} \tag{4.16}$$

The sign of quantity t is determined by the sign of the elevation of M above K. We can thus include in our consideration a new formal isoline, the found extreme point, and now perform the interpolation between the extreme isolines following the already described scheme and using this point. Note that in the proposed method the elliptic paraboloid is not used to approximate the relief function in a certain region but only as the basis for an heuristic formula in computer implementation.

Let us designate on the map the function thus constructed as $Z_1(p)$. Its obvious property is continuity, which distinguishes it favourably from the functions of unidimensional and network interpolation. However, at the points with an ambiguous shortest perpendicular to the isoline there will arise discontinuities of its first derivative. This can be avoided by substituting the distance from P to the corresponding isoline in the formula for $Z_1(P)$ by the averaged distance to the isoline from the points of a circumference of a small radius with the centre at P (Fig. 4.19). It is obvious that this procedure cannot be applied to the points of isolines themselves, since their elevations are fixed. The technique of constructing a new function, Z_2, described below, makes it possible to avoid discontinuities in the derivatives of Z_1, as well as along the isolines. Note, however, that in many cases it is sufficient to confine ourselves to the Z_1 function.

Construction of Z_2 begins with calculating the approximate values of gradients at an arbitrary point P on the isoline (Fig. 4.19). Let PM be perpendicular to the isoline at point P, $PM = \varepsilon$, $MT = K(\varepsilon)$ is a certain function of distance, then:

$$\text{grad } Z_1(P) = \left(\frac{H2 \times K(\varepsilon) + H3 \times \varepsilon}{K(\varepsilon) + \varepsilon}\right)' = \frac{H3 - H2}{PB} = \frac{h}{PB} = 0 \quad (4.17)$$

where h is the contour interval.

The value of the gradient, obtained while approximating point P from the other side, is equal to h/PA. We determine the gradient at P as the average:

$$\frac{1}{p + q}\left(p\,\frac{h}{PB} + q\,\frac{h}{PA}\right) \qquad (4.18)$$

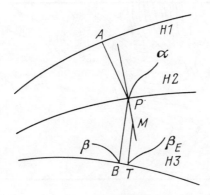

Figure 4.19

142

The weighting coefficients are chosen from considerations of convenience. They have three obvious pairs of values: 0.5 and 0.5; AP and PB; PB and AP.

We have plotted a gradient field along every isoline. The function along every isoline must have a preassigned value and a preassigned field of normal derivatives, which provides for the splicing of approximating functions by the gradient.

Construction of the Z_1 function at point P (Fig. 4.19) was based on linear interpolation along the curvilinear APB route. Construction of Z_2 is carried out along the same route, but linear interpolation is replaced by interpolation with a monotonous function obtained from the previously introduced $\psi_{\alpha,\beta}$ functions by scaling. Parameters α and β are prescribed by the corresponding gradient values. For instance, the value of the interpolating function at point M is found from formula

$$Z_2(M) = h\psi_{\alpha,\beta}\left(\frac{K(\varepsilon)}{\varepsilon + K(\varepsilon)}\right) \qquad \gamma = \beta_\varepsilon \frac{K(\varepsilon) + \varepsilon}{h} \qquad \delta = \alpha_\varepsilon \frac{K(\varepsilon) + \varepsilon}{h} \quad (4.19)$$

Having differentiated this expression with respect to ε at point $\varepsilon = 0$, we shall obtain α and β, that is the gradients at P and B, which proves the basic property of gradient continuity.

It is necessary to mention a certain drawback of using Z_2 at the above-mentioned points with an ambiguous minimal perpendicular to the isoline. A discontinuity of Z_2 occurs at these points. In most cases this discontinuity is very small, and when cross-sections and intermediate isolines are plotted it lies within the limits of graphical accuracy. In the general case, when constructing Z_2, one should average the gradients along the circumference to calculate the α and β derivatives, as was done in the above-mentioned operation of averaging the distance.

Let us now examine an example illustrating the use of Z_1 and Z_2 function for relief interpolation. Figure 4.20 shows a portion of a topographic map relief. After normalization it was inserted into an ES-1022 computer with the help of

Figure 4.20

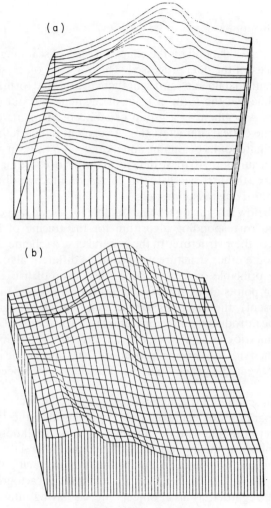

Figure 4.21

'Photomation' scanner. The model constructed by the computer was displayed on a 'Digigraph' graph plotter in the form of block diagrams formed by systems of profiles. Figure 4.21a shows the block diagram corresponding to the Z_1 function, Fig. 4.21b to the Z_2 function.

4.4. Tracing Linear Elements on a Map

Automatic reading (i.e. input into a computer and the identification) of cartographic data is a natural sequence of the following stages:

1. scanning and input into a computer;
2. data packing by lines;

3. tracing;
4. data identification.

The first two stages depend on the characteristics of scanning devices and not on the specificity of the data processed. The objective of the second stage is to receive the scanned data and represent them in a packed form, viz. line by line, and in each line—in the form of a set of segments cut out of the map by a scanning line. A segment is a group of adjacent elements of discretization having the same optical characteristics. A segment is thus defined by its class T and the coordinates of its beginning and its end—XB, XE. In addition to that a line has its own number. All these data form the list of coordinates for each class T. For instance, the first two lists can correspond to two different classes of markers, the third one to isolines, etc.

Examined below is a tracing scheme suitable for a wide range of applications, as well as the corresponding algorithm for the tracing of isolines and the identification of their structure. In the general case a scheme designed to trace data having some other structure will require a different algorithm. What can change is the principle of data packing, the set-up of lists enumerating the coordinates of points along the boundaries (in some cases these lists may be altogether absent), the functions for the computation of areas and distances may have to be introduced, or the operation of the withdrawal of objects on the basis of their metric characteristics, etc. The diversity of possible functions that can be assigned to the tracing procedure has thus made it expedient to distinguish between the general scheme of tracing and its specific implementation.

The general scheme of tracing presupposes that the lists of coordinates are processed separately for each class T. Assuming that the tracing has been completed up to the line with the number of M, we shall describe the typical tracing step following the input of the Mth and the $(M + 1)$th lines.

From the geometric point of view the function of tracing is as follows. All objects, irrespective of whether they are regarded in cartography as areal or linear, will be regarded as areal objects, i.e. as having internal points and a boundary. Each object will then be fully characterized by its boundaries (and its class). Tracing is assigned the task of identifying the boundaries of each object. The main requirement imposed in this case on the scheme for reasons of effectiveness (and when working in the 'on-line' mode this requirement is also dictated by equipment) is that all objects should be processed within one scan of the input data. In other words, each line is placed in RAM (of a limited volume) only once. Thus, in the course of tracing, only what has been derived from the previous lines can be used. A connected boundary is therefore processed by parts lying between the extreme points. When the tracing reaches these points the parts are spliced (parts AB, BC and the extreme point B in Fig. 4.22). Naturally, if the extreme point is that of the maximum, the two parts (AP, AB) will be spliced at the very beginning of tracing. It is, however, convenient to distinguish between these two parts, because with a fixed direction of outlining the boundaries (the region traced from the right), the lists

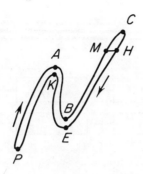

Figure 4.22

of coordinates for these parts are compiled differently: from the 'head' or from the 'tail'.

Every segment is involved in two tracing processes, going in different directions, corresponding to the two portions of the boundary (in Fig. 4.22, segment *MH*, parts *CB*, *CE*). The tracing is accomplished by the terminal points of a segment being assigned the reference values of the identified parameters (coordinates of the fixed nodes, the accumulated length, etc.). Each tracing step consists in correcting these parameters and transferring the references onto the next line, so that at the next step the parameters could be corrected and supplemented with the data for the new line.

Let digits 1, 2 designate the lines participating in tracing. All the segments of line 2 have to be examined in the course of tracing and their mutual location relative to the neighbouring segments in the same line and in line 1 established. The tracing scheme is shown in Fig. 4.23. The logic of the scheme is controlled by numbers $XB1$, $XE1$, $XB2$, $XE2$. The six individual blocks correspond to specialized subprograms whose implementation is not specified in the scheme. As seen from the figure, coming to the input of each block there are always two 'neighbouring' nodes. Their processing depends upon the type of proximity. The symbolic operator, 'NEXT', assigns to the parameters their corresponding values of the next segment.

Realization of the tracing scheme for normalized isolines is based on the following requirements:

1. the connectiveness component of every object on the map (isolines, markers, retouch) is represented in the form of a two-directional list of nodes (a node is a point on the map, prescribed by its coordinates);
2. the boundary is approximated by a polygon plotted on the nodes, so that a minimal number of nodes is required at a fixed accuracy of approximation;
3. after the tracing each list must be described by one descriptor.

The subprogram for the input and packing of every optical class forms each following line inserted into a computer as a set *E* whose element consists of the fields:

$$E: XB, XE, CB, CE, XCHB, XCHE$$

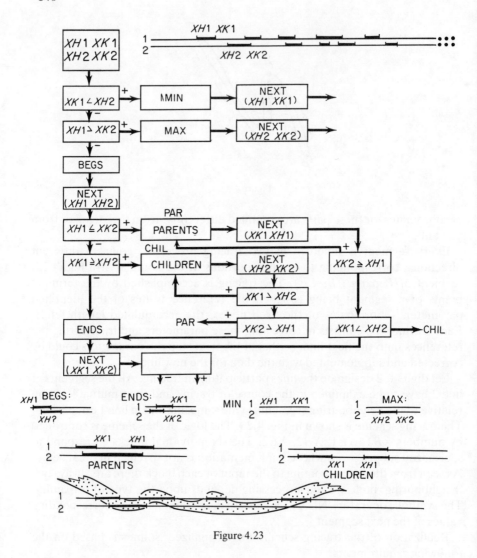

Figure 4.23

All the lists of nodes, irrespective of their class, are stored in one *MEM* file, whose typical element consists of the fields:

$$\text{MEM: } X, Y, CA, CB.$$

In the course of tracing, each portion lying between the extreme points has a corresponding descriptor with the elements:

$$\text{TAB: TYPE, } CM, CH, T1, T2, XB, YB$$

The functioning of this structure is actualized by utility programs, the most important of which are:

cyclic E-shift (there are two lines, $E1$ and $E2$);
issue a free MEM element;
issue a free TAB element;
connect the TAB element to the list of scratch;
organize the equivalence of two TAB elements;
connect the MEM element to the list of scratch.

The meaning of the above variables will later become clear from the context.

Every successive node is entered into MEM, and the CA and CB references are organized to match the subsequent and the previous nodes by a special program—KNOT. Its functioning will be described below, but at this stage it should be noted that it receives the 'candidate' for the node in a given list, which is either 'accepted' or 'rejected', depending on the state of the list. A special KNOTAB input 'accepts' the node without checking.

The BEGIN block. $XH2$ is a candidate for the tail of the list, prescribed by $XH1$.

The END block. $XK2$ is a candidate for the head of the list, prescribed by $XH1$.

The MIN block. Two lists, corresponding to XH1 and XK1, are joined in the sequence: XK1 → XH1, the XK1 and XH1 nodes are accepted, and the descriptors of these parts of the boundaries are made equivalent with the help of the CH reference.

The MAX block. Two nodes are set in the sequence: XH2 XK2. Two equivalent descriptors are identified with the help of the CH reference.

The PARENTS block is similar to the MIN block within the accuracy of orientation.

The CHILDREN block is similar to the MAX block.

The algorithm of the KNOT subprogram is illustrated by Fig. 4.24. Let us assume node A to have already been accepted and recorded. If M is the first candidate submitted for the next node, it can be recorded at once, without

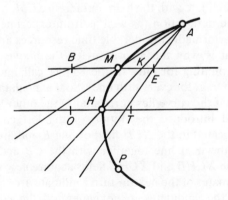

Figure 4.24

checking. A candidate is always submitted with a certain vicinity (in the given case, BE), used to formulate the criterion of the closeness of the curve and the polygon approximating it. If the subsequent edge of the polygon passes through this vicinity the criterion is regarded as fulfilled. The value for the vicinity is found as the minimum of two parameters: the absolute maximum permissible deviation of the curve from the polygon and a certain percentage of the distance to the neighbouring curve (isoline). Thus the denser the isolines in a given region or the greater their curvature, the closer the nodes.

Every next candidate for a node is approved not immediately, but after the following candidate has been checked. Let H be such a candidate. We draw a segment AH, intersecting the line prescribed by the BE segment at point K. If K belongs to BE, M is not accepted and the new candidate, H, is compared with the subsequent candidates. In this case the candidate following next must have its edge passing not only through the vicinity of BE of the first candidate, but also through the vicinity of OT of the second (in the general case, of any of the previous) candidate. In Fig. 4.24 the new point, P, is obviously lacking this property; that is why the AH edge is accepted, and point P becomes the first candidate for the next edge, etc.

The implementation of the algorithm would be very complicated if for every curve (and when the tracing is performed there may be many of them in one line) the vicinities of all the previous candidates were preserved. It is simpler to assign the vicinity not by its size but by a pair of angles formed by the respective half-lines with the vertical line (e.g. the angles between the vertical line and the BA and BE half-lines). Then point H is accepted if the angle for AH lies between the angles of the vicinity. When accepting point M, we choose (from the two half-lines, BA and OA) the one nearest to point H, and regard its angle as one of the angles of the new vicinity. The second angle is found in a similar way.

Coordinates of the last point to be accepted are stored in the TAB descriptor (it is referred to by the element of line $E1$ with the help of reference C): XB, YB. The angles of the vicinity, $T1$, $T2$, are also kept there. The descriptor refers to the previous candidate with the help of reference CM.

From the above description (as well as from experiments) it is seen that, when tracing is performed, considerable time resources are spent on this very subprogram. That is why the algorithm for the above BEGIN and END subprograms, accounting for most of the KNOT calls, generates candidates with a certain step (e.g. for every fifth line). So that no data are lost in this case (which is possible if the curve lies horizontally) the other lines are checked as follows. We shall introduce the notion of the last checked node whose coordinates are located in the XCH field of line E and are either transferred unchanged onto the next line together with the CB and CE references or redetermined. The $XCHB$ and $XCHE$ fields acquire new values in two cases. Firstly, the coordinates of the next-in-turn candidate are sent there. Secondly, in the lines where the candidates are not generated, the values for XB–$XCHB$ and XE–$XCHE$ are checked. If the absolute value of one of them exceeds the permissible limit, a candidate is formed, and, consequently, the last-checked node is memorized. Note that this check can be performed both in the course of tracing and after it.

The stage of the traced data identification consists in restoring the original structure of isolines: returning from the 'double' representation by points lying on the boundary of the isoline treated as an areal object to its linear representation; eliminating the discontinuities; and restoring the isolines concealed by retouching. These problems can be solved by using a special set of functions (subprograms) for the processing of graphical data. The whole set is in fact a means of processing any kind of scanned normalized maps. Some typical subprograms used for isoline processing are described below.

Function 1: moving along a contour, obtained as a result of tracing with an arbitrary step

Let us examine a certain contour (Fig. 4.25). We assume to have been given a step of movements, T, along this contour, and we are now at point B. If $BE > T$ the subsequent point C lies on the same edge as point B. Its coordinates are easily found if we know the direction cosines of the AE edge (and of course the coordinates of point B as well). How can we find the next point, M, if $CE < T$? The remainder of CE and all the 'intermediate edges' (EK) should be traversed in search for the KO edge satisfying the condition of $KM < KO$ with the sum of $CE + EK + KM$ equal to T. The variables of the subprogram for an arbitrary point M will thus be represented by the number of the edge, the coordinates of M, the distances to the MK and MO vertices (if it is necessary to move in both directions) and the direction cosines of the KO edge.

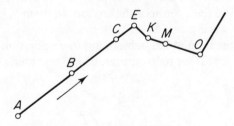

Figure 4.25

Function 2: distinguishing the terminal points of linear objects

Two methods for the realization of this function will be proposed. Movement is always understood as function 1.

The first method Two empirically determined numbers are preassigned: E the step of movement, K the number of E steps between two movement processes. The method is illustrated by Fig. 4.26. Let us assume that arbitrary

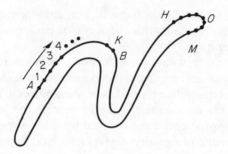

Figure 4.26

initial states of the above two processes are A and B, make these processes follow the same direction, and introduce a third empirical parameter L. If in the course of movement the distance between the points becomes less than this number L, the terminal point is regarded as having been identified (positions of points in the figure—N and M).

The second method   We first find the longest edge of the contour (AB in Fig. 4.27). Let M be its median point. A certain point P will move from M in a certain direction with the assigned small step until the distance between M and P becomes sufficiently small. Now we can use function 3 with the initial state of M and P. It may so happen, however, that the above-mentioned point P will not be found (e.g. if there is noise in the vicinity of M on one of the 'banks' of the isoline; incidentally, the longest edge has in fact been chosen to decrease the probability of this happening). If this does happen then another long edge should be taken instead of AB.

If, using one of these methods, the terminal points are not found, it is inferred that they do not exist. We then consider the contour to bound not a linear but an areal object.

If, however, the terminal point has been found, the operation of specifying its coordinates is called for. This procedure can also be regarded as elementary.

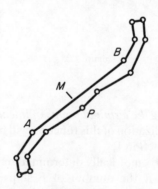

Figure 4.27

We shall now briefly describe it. We assume that the terminal point has been found in position N, M (Fig. 4.26). We then reduce the step of movement several times (e.g. ten-fold) and make points N and M move towards each other so that with every step the distance between them would either decrease or have the minimum possible increment. 'Balanced' movement is achieved in this way, and with the bending of the end its comparatively accurate position is obtained. It is point O, where N and M merge.

Function 3: movement along a linear object in search for either the terminal point or a noise
It should be noted that this function, and the two functions following it, can be realized in two ways. They are characterized by different parameters: time, complexity, accuracy, etc. The choice of the method depends on specific requirements.

The first method Let us assume that we have chosen two points on the opposite banks of an isoline with the distance between them being sufficiently small (A and B in Fig. 4.28). We then select a small step E. It is found empirically and should invariably be less than the typical width of the isoline. The next position of this concordantly moving pair of points can be obtained if one considers the E shifts of each one of them. Out of three pairs, A, $E(B)$; B, $E(A)$; $E(A)$, $E(B)$, the one that determines the shortest distance is chosen.

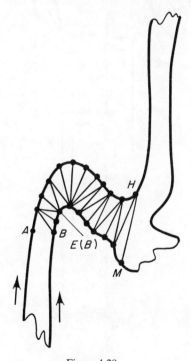

Figure 4.28

Subsequent movement proceeds similarly, and each time the length of the chosen segment is compared with a certain value. When this threshold is exceeded a noise (*MH*) is considered to have been encountered.

The second method   It is easy to see that the above method is effective only if the step of movement is small. On the one hand, it slows down the programme; on the other hand, the situation of dealing with noise arises frequently. Noise has to be processed by special means (function 4) which should be resorted to only when it is really necessary. In particular, it is desirable to 'pass by' the small noises.

The second method satisfies these requirements. As above, let *A* and *B* be taken as the initial points. Both the step of movement and the criterion of length *E*, when the next-in-turn pair of points is accepted, can be larger than in the first method. Figure 4.29 shows the directions of shifts *P* and *N*. Point *A* is the 'leading' one in this process. Small noises are 'passed by'. The size of the step

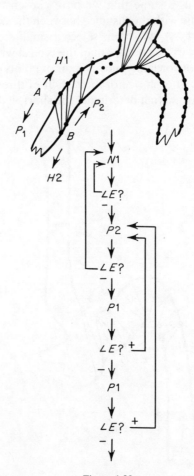

Figure 4.29

and of the maximum segment depends on the maximum noise along the boundary, which does not have to be processed.

It only remains to say a few words about the search for the end of a linear object while moving along it. We first determine the length of that portion of the contour along which the movement is taking place. Each subsequent step decreases this length. As soon as the remainder becomes smaller than permissible, the terminal point should be processed. The algorithm has already been described (see function 2).

Function 4: elimination of noise at the boundary of a linear object
In moving along a line, once noise is encountered it should somehow be taken into account and processed. The first step is to replace the 'critical' position of the pair of points (A, B in Fig. 4.30) by their regular position before the noise. This is done by moving backwards with a small step, fixing the 'narrowest' position (N, P). Then, with the same small step, points are placed along the boundary in the forward direction. The number of these points serves as the initial parameter and depends on the maximum permissible noise. This is followed by checking the distances between the points for each pair on the opposite banks. As soon as a distance smaller than a certain threshold is found, it is fixed as the end of the noise (to be more precise, first, as in the previous case, the 'regular' position of the end of the noise is searched for: M, T in Fig. 4.30). The physical removal of noise depends on a concrete data base and will not be dealt with here. We shall only indicate one possible way: the centres of the segments lying before and after the noise can be connected (Fig. 4.30).

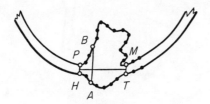

Figure 4.30

Function 5: thinning the linear object—distinguishing its median line
The purpose of the respective subprogram is to dispose of the redundant points not needed for subsequent processing by replacing two banks by one line regarded as more 'plausible'. As above, more than one realization route is possible. Both the peculiarities of the problem and the computer resources should be taken into consideration.

The first method The first algorithm of moving along a linear object can simply be supplemented with a block fixing the centres of segments. These centres will constitute the median line sought. The disadvantage of this approach is that a large number of nodes are needed, as the step is small. If the second method is taken for the realization of function 5, a distorted median line may be obtained. This situation can be rectified if, as in the previously described algorithm for the placement of nodes, only some of the nodes are 'accepted'.

The second method This can be used to advantage in constructing digital relief models, if one bears in mind the fact that the position of isolines themselves on a topographic map is determined only with a certain accuracy depending on the contour interval. It is suggested that after finding the terminal points and isolating the noises we simply restrict ourselves to one of the isoline banks. As a variant, both banks can be left after first having been separated; this, however, will slow down the programme of constructing the digital model.

The third method This method seems to be the most appropriate in digital models. Both banks are first identified. One of them is referred to as the leading one (*A*, Fig. 4.31). The coordinates of each of its nodes are corrected in accordance with the following rule. Let *P* be the node, and *PB* and *PC* be the edges adjacent to it. Bisecting the *BPC* angle we find the intersection point *O* on the opposite bank. The centre *OE* is the new position of *P*. It only remains to connect the points obtained.

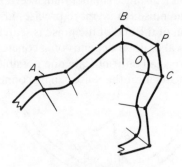

Figure 4.31

4.5. The Structure and Functions of the Bank of Spatial–Digital and Cartographic Data

Purpose

To provide a possibility for the simultaneous processing of spatially distributed data: cartographic, statistical, aerospace photographic.

Structure

The software for the bank includes the following basic components:

1. a system for the automatic readout of data from maps, photographs, and other graphical documents and their conversion into the bank formats;

2. a system dealing with map projections and photograph distortions and referring the data to a standard system of coordinates in the bank;
3. a database management (operating) system;
4. a request–reply (interactive) system providing for a dialogue between the user and the bank;
5. a system of graphical visualization (including the plotting of attendant maps);
6. a system of duplicating the bank data on normalized maps;
7. a system of creating and decentralizing small thematic (specialized) data banks based on normalized maps;
8. a package of special application programmes.

Data structure

A data element in the bank is a function defined on a certain subset of the coordinate plane. The functions are classified by the dimensionality of their domain (i.e. the set where they are not trivial) and by the set within which the function acquires specific values. A range of values is always represented by the direct product of the Euclidean space and a certain finite set (qualitative parameters): $R^N \oplus M$. Depending on dimensionality of domain the functions are subdivided into point, net, and areal ones.

Database structure

Each of the above-mentioned types of functions prescribes a certain type of file designed for the storage of the descriptions and data related to this function. A set of function types in their totality defines the database.

A file of points is a list of their coordinates at which the value of the function has been defined, the range within which these data lie, and the values themselves. As an example: populated localities of the Moscow Region, the values—population, industry, distance from Moscow.

A net file is defined as a file of points (vertices) plus edges (possibly with orientation), on which a function can be determined. As an example: a drainage network, with the edges representing the median line of a river channel and the function the width of a river and other linearly distributed parameters.

An areal file has several varieties, depending on the range of values and certain additional requirements (accessibility, accuracy, memory parameters, type of problem solved, ease of data modifications, etc.). These varieties are described below.

1. Range of values—a finite set
This function divides a plane (or its subset) into a number of non-overlapping areas, and has a constant value on each of them. Three organizational types are proposed for this file. The first one is that each area is represented in computer

memory by its internal boundaries. To be more precise, an area has a corresponding polygon approximating it on the plane and a corresponding value. This information is to a certain extent redundant, since the neighbouring areas have part of the boundaries in common. The advantage of the method lies in the simplicity of obtaining a file automatically in the course of scanning and data tracing, in the ready accessibility of the basic information—the area. A file thus organized is called the areal file of boundaries.

Another type of organization is that of a file of oriented edges. In this case the redundancy inherent in the previous type is eliminated by distinguishing the median lines and branch points. Each edge is oriented at random. The orientation of the plane induces a standard orientation on each area. Comparison of orientations in processing the file makes it possible to determine the area to which a certain edge belongs. A file with such an organization presents the data in a compact form and is effective when the methods of access are based on finding the neighbouring areas. This file has a certain resemblance to the network file. The main difference is that here the data are associated with an area and in network files they are localized on a network.

The third type of areal file is a file of sections. In this type the data on the boundary of every area are not stored. Instead, sections of the domain of definition are plotted with a certain step, parallel to a certain direction. The section of an area is stored as three numbers: the entry point, the exit point, and the number in the catalogue of areas. The catalogue itself assigns certain thematic characteristics to every area. Such organization permits easy transition to matrix representation (see below), the access to which, if a high accuracy is not required, is very prompt and convenient.

2. Range of values—real numbers

Here, too, three varieties of files are proposed. The first one is a file of isolines, based on direct storage of the isolines in a certain inner representation (usually in the form of polygonal lines) within the limits of the domain of definition. An isoline has a corresponding elevation assigned to it. Several modifications exist for this file, associated with its possible applications. The niceties (the structure of redundant data) are omitted. The main purpose is the construction of digital relief models and visualization of contour lines.

The second variety is an analytical function assigned in the form of a bicubic spline on a rectangular grid. Such a function is unambiguously assigned by four numbers on the grid nodes. The advantages of the structure are simplicity of organization and software, feasibility of being used to solve the problems requiring differential calculus and for the drawing of intermediate contours.

The third variety is as a rule used for the functions prescribing the relief. An area is divided into curvilinear segments (quadrangular), whose vertices represent the critical points of a function: one minimum, two saddles, and one maximum. A segment is often present in truncated form. The edges of a segment represent the water-parting lines. This kind of file is used to solve the problems associated with finding the boundaries of water catchments, etc.

3. Range of values—the direct product of $R^N \oplus M$

It is proposed that the data of this file should be stored in the matrix form. A certain rectangular grid is prescribed, with the data projected onto its nodes. An obvious advantage of the structure, along with its simplicity, is the possibility of its being effectively corrected.

The operating system envisages a means of declaring a certain file to be dynamic. This means that the objects described by this file are subject to certain changes, and it is desirable to fix all the intermediate states in the bank. In some cases such a declaration has no sense (e.g. for contour lines); in some cases the history of the development of an object is fixed; in some other cases there are seasonal fluctuations (floods, meandering of river channels, etc.), which make it possible to ascertain the standard state of objects. The simplest method for the realization of dynamic files is to store all the temporal 'cuts'. An enormous redundancy of information is, however, obvious here. An apparatus for the editing of the structure is envisaged for such files. Each temporal level is represented by a set of the edited previous levels. For some levels the original file is stored, which makes it possible to avoid prolonged reiterative editing. A dynamic file is thus, in the local sense, a combination of two files: the master file and the file of editions associated with it.

4.5.1. The function of normalized maps in data banks

The data bank for a certain territory can be regarded as satisfying the necessary requirements if: digital models have been constructed for this territory; all the models related to different territories are conformably spliced; excessive redundancy has been eliminated in all of them; there exists a means for concerted control of these models and for the control of visualization data. The volume and complexity of such a data bank depend on the nature of the problems being solved. These can include such complex interrelated problems as supplying information for the management of an industry, forecasting the development of economic and natural resources, processing the data of tele-probing, or map plotting.

These banks are quite voluminous and complex. This is why, for economic reasons, they can only be used by large organizations. If the problems to be solved using this data bank are only those of providing information for such large organizations, the effectiveness and value of such a bank will be low, and the desirable promptness will not be achieved. The possibilities of a normalized map turn out to be useful here. The means of visualization provided in the bank make it possible to compose automatically and promptly series of fully normalized maps. Moreover, the bank is a suitable tool for designing and adjusting different normalization methods. An important feature of such series of maps will be their full conformity to each other, as well as to the data in the bank. Each map carries coded information on its purpose and spatial fixation. No difficulties arise when the automatically processed maps of these series, representing different territories, are joined together. These maps are provided with graphical redundancy on the margins, making it possible to devise the means of data input control assuring an acceptable degree of identity between the input data and their prototypes in the bank.

We thus see that such normalized maps perform an additional function: taken together they constitute graphical copies of the master data bank. They are a tool for the deconcentration and decentralization of the data bank. It would be possible to use the bank in small computing centres and even in the field, if a reliable mini- or microcomputer, a scanner, and the necessary series of such normalized maps are available.

The system described above operates at two levels. The upper level—a classical data bank on a large computer with capacious memory. The lower level—a copy of part of data on paper in the form of normalized maps for information retrieval and processing by man.

The data banks of these two levels are designed to solve different problems. The master data bank is assigned the functions of constructing and updating the normalized maps and the catalogues for file control systems of local banks. The dialogue with a large data bank can be more complicated. The master data bank processes the data obtained from remote sources and from maps. If the newly arrived data are considered reliable and valuable for many users, they are processed and printed in the normalized form to be sent to remote users (and included in a normalized data bank of the lower level).

Small data banks on mini- and microcomputers are not used in the processing of data from remote sources. Their objectives are as a rule relatively more modest: to process the available normalized maps, obtain derived data, construct derived maps, block diagrams, etc. Specifically, such is the case with the automation of cartometric works in highway engineering or water-development projects, when it is necessary to find optimal routes, distances, volumes, depths, etc. A request arrives as a text and some normalized maps placed into the reading device. A request may be sent in a graphical form: as additional objects on a map, an approximate zone of search, an approximate solution, the position of a dam on a river, or a digital code sign distinguishing the area of interest. If additional maps prove to be necessary to solve the problem, the system will automatically make a request for them using special addresses or names, and they will be entered in the assigned sequence by the operator.

The systems based on the above data banks produce either digital or graphical data. In the graphical form attendant maps can be produced to supply cartographic data for prompt representation of a certain state and a specific forecast (e.g. a map of soil moisture content, a map of soil moisture forecast). The language of representation here is maximally simplified for the promptitude of map plotting.

CHAPTER 5
Automated Map Compilation

Automated map compilation has been extensively developed in recent years. The greatest progress has been made in compiling maps based on digital data. This is explained by the simpler algorithmic presentation of the processes of machine-assisted processing and compilation of maps. Compilation of maps on the basis of original cartographic data appears to be a much more complicated problem. As yet not enough research and development work has been carried out to result in complete automation of all map compilation processes. The main difficulty here is known to lie in automatic identification of cartographic symbols by a machine. That is why there is now a tendency to develop semi-automatic methods of map compilation, using interactive dialogue devices. Interactive processing is especially effective in the editing of maps. The main objective, however, is still complete automation of all processes of designing and composing a smooth-delineation map. The present chapter deals with the methods of achieving complete automation of the map compilation processes, based on the principles of normalization and scale generalization. The main emphasis in these methods is placed on the accuracy and the detailedness of data representation.

5.1. Automatic Map Compilation Based on Normalized Field Survey Sheets and Deciphered Aerospace Photographs

To assure total automation of the process of map compilation from deciphered aerospace photographs, field survey sheets, or any other cartographic material, it is necessary that these should be represented in a normalized form. One of the methods for the automatic processing of a normalized image of areal objects presented as contours, further referred to as areas, and compiling maps with the help of a computer, scanner, and graph plotter [124] is described below.

From the formal point of view, a map of areas represents a set of areas specified by their boundaries. Such a map is processed by entering it into a computer and composing suitable description of the input data, which makes it possible to outline the boundaries with the help of a graph plotter (if necessary, in a smoothed-out form), drawing within the boundaries of every area a certain system of rasters, depending on the characteristics of this area, determining the areas and the lengths of boundaries, combining some areas into larger ones with respect to a specific parameter, etc.

It is assumed that the boundaries of areas are presented on the map with a colour (preferably black) that can be easily singled out by the scanning device and the computer against the background of the other substantive elements of the map. The objective of normalizing such maps is to designate some (or all) the areas with special digital code signs (DCS). These signs make it possible to establish the interrelation between the computer and the cartographer: with the help of DCS it is possible to interrelate any (quantitative and qualitative) information characterizing an area, to designate in the same way certain equivalent in some respect areas which are (e.g. forests), or set the task of calculating the areas and to obtain at the output the table of the areas found together with the corresponding DCSs. If the compiled map has a unified system of designations for background symbols (such as those of geological and soil maps), a unified system should be composed in the form of a standard table. For an arbitrary system of background symbols, i.e. when the cartographer designs his own system of background symbols, a certain degree of freedom in the choice of parameters and combination of lines is assumed, but it is expedient to take them in accordance with the parameters of the grids of background symbols for the designed map. Thicker lines are admissible to facilitate the process of drawing the background code symbols. In an overwhelming majority of cases, to compose the whole map legend it is sufficient to use three different thicknesses of grid lines and three or four colours. From the law of subtractive and spatial mixing of colours it is known that by combining linear grids and colour fillings with grids consisting of three basic colours (blue, purple, yellow), one can obtain all the other colours of the colour circle.

Let us examine the sequence of fully automatic operations aimed at creating a thematic contour map, starting with scanning the normalized source map or the results of deciphering, and ending with a colour-separation final compilation or a multicolour map obtained on a graph plotter.

1. Scanning with a photoelectronic device quantizing the image by brightness levels, with the data represented in computer memory in the form of digital matrix p. Each of its elements, p_{ij}, is the code of the quantization level at points uniformly distributed over the map, where i is the number of a point in a line and j is the number of the scanning line.

2. Processing the input data. This includes the following operations:

   (a) identifying the digital code signs and the points of contour boundaries;

   (b) tracing the contour boundaries, i.e. arranging the coordinates of points on these boundaries in an ordered sequence;

(c) correlating the digital code signs with the numbers of the map legend signs stored in computer memory.
3. Plotting the map image. If the map is to be printed in large numbers it is expedient to engrave the originals separately, according to the colours of signs in the legend. If the map is not intended for duplication, the multi-coloured image is plotted immediately.

The principles of raster discretization are used to simplify the procedure of composing a map with the help of a graph plotter in the form of qualitative background (e.g. on geological, soil, land use, and other maps), as well as to assure a high reliability of the identification of signs in subsequent automatic reading of the map. Only the specialized part of the image is automatically processed and plotted, i.e. the information determining the specific objective of the map and applied onto a standard base. For the maps of the above type, specialized elements of the image are the background (filling the contours), representing the qualitative characteristics of objects, and the boundary lines of contours.

Let us now examine the procedure of applying rasters on an area. Raster combinations are conveniently stored as a set of numbers: $N_1, W_1, N_2, W_2, \ldots,$ N_k, W_k. Here k is the number of lines in one raster combination, N_i is the pen thickness (if necessary, also the colour), W_i is the distance between the axis of the ith line and that of the neighbouring line, W_k is the distance between adjacent combinations. Raster combinations are stored in computer memory in the form of such vectors. A combination for every area can be either prepared or taken from the file.

Let us assume that for a given specific area a certain combination of rasters has been chosen. The structure of data obtained as a result of tracing (a set of branch points and the arcs connecting them) makes it possible to represent the boundary of the area in the form of a sequence of points. A raster, however, has to be represented in a special 'raster' form. It is convenient to presume that such a combination of rasters is plotted over the whole field of the map, the graph plotter pen being lowered inside the area and raised outside it. The coordinates of the points where the pen is either raised or lowered make up the database in the 'raster' form. This structure can be obtained in the following way. Let us examine segment OE (Fig. 5.1) determining the next-in-turn raster line. It is necessary to find where it crosses the boundary of the area. Point O can be assumed to lie outside the area, whose boundary has been approximated with the necessary degree of accuracy by a polygon. For each of its sides we check whether it intersects the OE segment. All the intersection points are ordered from top to bottom ($a, b, c, d, \ldots$ in Fig. 5.1). It should be noted that the points of tangency, very easily identified in polygons, enter this set twice. The sought raster line segments (ab, cd, ef, fg, hi) are obtained if we connect the first point of this set with the second, the third with the fourth, and so on, or, what is the same, the points where the pen is lowered (a, c, e, f, h) and where it is raised (b, d, f, g, i).

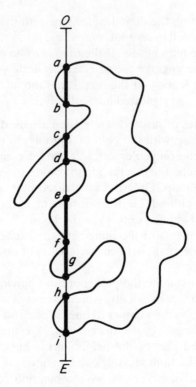

Figure 5.1

5.2. Automatic Representation and Reading of Discrete Data on a Map of Populated Localities

The technique of composing normalized maps by the DCS (digital code sign) method, as well as an algorithm of reading such maps with the help of a scanning device and a computer, is described below. A special program was developed to place the initial data in computer memory, process them and produce a derivative map on a graph plotter (Fig. 5.2).

The specialized content of the map is represented by raster code signs consisting of strokes of a differing thickness. Stroke thickness carries a certain weight: 1000, 100, and 10 inhabitants for the map of 1 : 500,000 scale. Combinations of these strokes (and half-strokes) characterize the population of specific localities with an accuracy of 2.5 persons. The lower left corner of the extreme left stroke in every sign is located in the centre of the populated locality. The method in question has a number of advantages:

1. prompt automatic plotting and identification of signs;
2. ease of the visual read-out of signs;
3. compactness of a sign, which enables one to enlarge the information content of a map; thus, it was possible to retain all the populated localities on the normalized map.

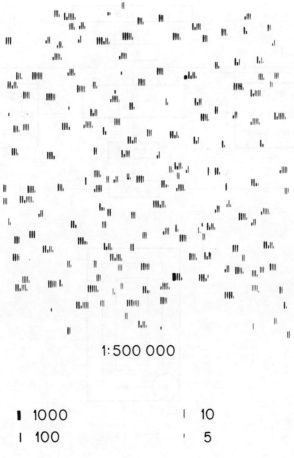

1:500 000

▌ 1000		❙ 10
❙ 100		' 5

Figure 5.2

The block diagram of compiling the specialized content is shown in Fig. 5.3. The initial data for the program are the coordinates of a populated locality (x, y) and the number of inhabitants B. These data are either retrieved from the files of a cartographic data bank, or put in with the help of a digitizer directly from the cartographic material. The chosen weights (in our case, 1000, 100, 10) are placed in the A massif. Subroutine CC applies an individual stroke onto the map. Its parameters are: (x, y), the coordinates of the lower left corner of the stroke; N and K, the type of a stroke by height (1 or 2) and thickness (1, 2, or 3). Concrete values of heights and thicknesses are assigned as the initial data (e.g. thickness is placed in the W massif). The path of the graph plotter pen, when drawing the stroke, depends on the relationship between the thicknesses of the stroke and of the pen. An example of such a path is shown in Fig. 5.4. The initial point of a path has coordinates $(x + M, y + M)$, if $2M$ is the thickness of the graph plotter pen.

164

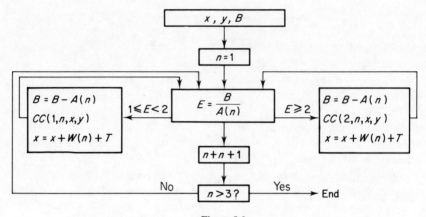

Figure 5.3

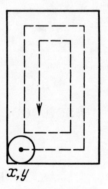

Figure 5.4

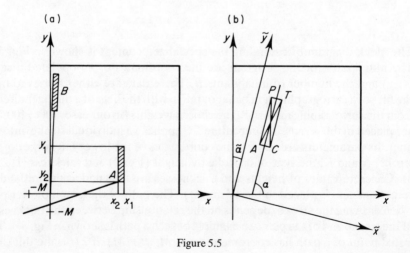

Figure 5.5

The specialized content of the obtained normalized map can be placed in the computer with the help of a scanning device. The algorithm for input data identification is rather simple. Scanning is assumed to be performed perpendicular to the strokes.

At the first stage each stroke is traced and processed independently. In the given simple case any other tracing program can be used, describing a stroke with a sequence of its boundary points. Then, using the obtained sequence, we find: area S, height $H = y1 - y2$, and point A lying the nearest to a certain point $(-M, -M)$. This point A is regarded as indicating the lower left corner of the stroke (Fig. 5.5). The stroke thickness is found from S and H.

It should be noted that in actual practice it is difficult to position the map on the drum of a reading device precisely enough for the scanning head to move exactly perpendicular to the strokes. Some errors are always possible. That is why stroke thickness is not immediately found as $x1 - x2$ but through the height, which is more accurate. With the same purpose we take the $(-M, -M)$ point and not the origin of coordinates: with a small tilt not the lower left corner but a certain point of the base (e.g. stroke B, Fig. 5.5a) may turn out to be the nearest.

A program has been developed that makes it possible to find and compensate for small accidental displacements of a map. When examining each stroke it is first necessary to find corner points A, P, T, C as the points nearest to the vertices of a certain large square, centrally located in coordinate system (x, y) with a side equal to $2M$. For instance point C, the nearest to $(-M, -M)$. The corrected thickness value is found as the arithmetic mean:

$$W = \frac{1}{4}\left(|AC| + |PT| + \frac{S}{|PA|} + \frac{S}{|AC|} \right) \tag{5.1}$$

and the tilt, α, as the mean of the tilt angles PC and AT. By averaging α relative to all the strokes, we shall find the rotation angle, $\bar{\alpha}$, for the coordinate system of the map relative to that of the scanning device (Fig. 5.5b). After that all the coordinates are expressed in the $(\bar{x}, \bar{y})$ coordinate system.

The type of a stroke by its thickness and height can be identified in two different ways. One possibility is to prescribe permissible values on the thickness and height scales. The number of the interval within which the concrete values W and H fall will then be the type sought. On the other hand, we can only assign the value of permissible scattering of the W and H values (i.e. the spacing). The location of permissible intervals on the scales is then found automatically. It is to be noted that the second method is more convenient, since the scattering value is standard: it depends on the accuracy of the utilized type of normalized maps and the accuracy of the algorithm, whereas a scale of thickness heights depends in each particular case on the specific normalized map.

The next stage is the 'assembly' of strokes into individual symbols. The feature used here is very simple: the y-coordinates of the lower left corners almost coincide, and the distances between the strokes, if ordered along the

x-axis, are constant (for all symbols). The coordinates of the lower left angle of the extreme left stroke of the assembled symbol will indicate the centre of the populated locality. Knowing the type of each stroke (and, naturally, the preassigned weights) it is easy to find the total number of population.

5.3. Automatic Data Representation by the Method of Multiparametric Continuous Cartogram

The technique in question has been developed based on the theory underlying the MRD method of multiparametric continuous cartogram, described in Chapter 2. Maps of such type are compiled from statistical data and a map of the boundaries of regions. In this case a region is understood as any territorial unit (e.g. an enterprise, farm, or district) for which statistical data are available. To provide for a higher level of automating the initial data input and map plotting it is expedient that the boundaries of the regions on the initial map should be given in a colour (preferably black) that can be reliably distinguished from the overall image of the map by a computer. It is even better to use a schematic (diagrammatic) map showing only the boundaries of regions. The input of the initial data and the correlation of regions by names or numbers with the statistical data can be accomplished in different ways:

1. The initial map is entered into a computer by a scanning device. The boundaries of regions are traced from the memory; that is the ordered sequence of coordinates is determined for the inner edge points of the boundary line. In the computer—from the program, and on the map—visually, the extreme upper point is found for each area in the course of scanning from left to right and from top to bottom; the numbers of the regions are assigned in the same order. These numbers are correlated with the appropriate statistical data. The subsequent procedure consists in the input of statistical data into the computer and the plotting of the map on the graph plotter.
2. Every region of the initial map is marked with digital code signs (DCSs) placed in the middle of it. The numerical DCS values are assigned corresponding to the numbers of the regions in the list of statistical data. Subsequent operations, performed automatically, are: the input of the map by a scanner into the computer; tracing boundaries; identifying the DCSs; the input of statistical data, and map plotting. The algorithm is described in detail in reference 144.
3. The initial map of the regions is inserted into a computer. The boundaries of the regions are traced. The results of tracing are shown on a graphical display. The correspondence between the regions and the respective statistical data is established with the help of a graphical dialogue system. This is followed by the plotting of the map.

Each of the techniques has its advantages and shortcomings. The first two have an essential advantage of not requiring the use of a graphical dialogue system.

No special equipment is required for the preparation of the map manuscript; only a pen or a pencil is needed. This means that initial maps can be prepared manually and then promptly processed automatically. The second method is more reliable than the first one since the regions are identified more accurately, but it is inferior to the third method as it requires higher labour expenditures.

Presented below is the algorithm for the preparation of a map following the second method (described in reference 144), starting from the moment when the input of initial data and their identification have been performed.

Let $P_1, P_2, \ldots, P_N$ be the densities of objects in individual regions ordered in a rising sequence, L_i is the largest possible distance between the axial lines of the raster in the ith region, e is the minimum permissible distance between two adjacent raster lines. When choosing the weights we take into account the resolving power of the reading device, raster sharpness, and the required accuracy of data representation. It is necessary to subdivide all regions into groups, to assign each group with a definite weight p_k (k is the number of the weight), and to determine for any region the distance l_i between the axial lines of the raster (i is the number of the region). Note that since the weight, p, is designated on the map by raster line thickness, T, there is a functional dependence between these two parameters. We shall designate the raster line thickness for weight p_k as T_k.

Let at first $p_1 = L_1 P_1$. This weight is characterized by the distance between raster lines in the first region being the largest ($l_i = L_1$). We assume that in the regions with numbers $2, 3, \ldots, i - 1$ the information is represented by the p_1 weight i.e. $T_1 + e \leq l_i \leq L_j$ ($1 \leq j \leq i - 1$). If $(T_1 + e)P_2 \leq P_1 \leq L_i P_i$, then weight p_1 is also used for the ith region. If $L_i P_i < p_1$, weight p_1 is corrected: $p_1 = L_i P_i$. The case of $(T_1 + e)P_i \geq P_1$ indicates that the possibilities of weight p_1 have been exhausted. To represent information in the regions with the numbers larger than $i - 1$, weight $p_2 = L_i P_i$ is introduced. The above procedure is repeated for it. Then weight p_3 is chosen and this continues until all the regions have been covered.

For a fixed region a raster consists of equally spaced segments of equal thickness. Thus each such raster is characterized by a pair of numbers (T, l), where T is the thickness of the segments for the given region; l is the distance between the axial lines of the raster. The T value depends on weight p chosen for a given region. The concrete form of this functional dependence is either inserted into a computer as initial data or should be determined by means of a program. Note that the automation of this process is closely associated with the problem of choosing the weights. The relationship between T and p is discussed in reference 121 and in section 2.4.1.

The gist of the subroutine for the output of segments is as follows. A file is scanned, record by record. Simultaneously the graph plotter pen moves along the map, duplicating the corresponding movements of the reading head. Where should the pen be lowered to draw a segment? Let each successively retrieved element of the massif have the form of (XL, XR, IN). Ordinal number IN determines the DCS of the corresponding region and, consequently the raster (T, l). It now remains for us to examine the value of $M = \mod (N, l)$ —the remainder of dividing the number of the map line, where the examined segment lies, by the period of the raster, l. If $M < T$ the segment is plotted. Further details are omitted because of their routineness.

5.4. Automatic Compilation of Maps by the Method of Multiscale Representation of Areal Objects

The method of multiscale data representation on maps was discussed in section 2.5.2. In this section we shall present the algorithm of automatic map compilation by this method, as applied to areal (contour) objects [110].

Let us assume that on a large-scale source material the objects are represented in the form of closed, arbitrarily distributed contours, which may exist in combinations with other elements of the image, indicated by different colours (Fig. 2.17a). The problem is to compose a map to a smaller scale representing the area, localization and, whenever possible, the shape of the contours (Fig. 2.17b).

Let us now consider the technique of automatically composing multiscale raster maps in compliance with the above requirements.

The process of map preparation is divided into three basic operations:

1. automatic data readout and input into a computer with the help of specialized photoelectronic devices;
2. data processing by the computer;
3. data registration at the computer output with a graph plotter, an alphanumerical printer, a cathode-ray tube, or other facilities.

The first operation is performed by a reading device with linear scanning. In the experiments in question a scanner was used that read out not only the elements of the map image but also the blank field of the map. The readout was performed point by point with a constant recurrence of points.

At the computer input the points were successively recorded in RAM bits, so that each bit represented a certain point. Such a form of recording is expedient when an image has two gradations (lines of the same colour on a white background), i.e. the blank field of the map is 0, and the content is 1.

In those cases, when contours are represented in the form of lines or a background with several brightness gradations, and selective readout of contours is needed, i.e. selection by brightness, the image is quantized into several brightness levels and some bits are assigned to code the brightness gradation.

The remaining data processing procedure consists of the following steps:

1. selection of the coordinates of points by contours;
2. computation of areas for every individual contour;
3. grouping the objects (contours) by values of areas in accordance with the taken scale factors;
4. determination of the centres of gravity of contours;

5. transformation of contours into the multiscale raster form of data representation, with allowance for the scale factors and the data written out by a
   graph plotter.

Selection of the coordinates of points by contours stored in computer memory
is a complex procedure. Before passing on to it let us examine the algorithm of
composing a multiscale raster image:

1. For every processed line the left and right extreme points of each contour
   are singled out. In Fig. 5.6 these are A_1B_1, A_2B_2, etc.

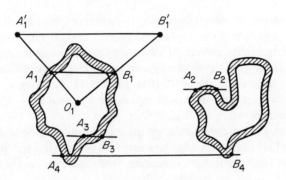

Figure 5.6

2. From the coordinates of points on the initial map contours we compute the
   coordinates of the points of the contours to be converted into other scales,
   in accordance with the accepted classification by dimensions of contour
   areas. For instance, from the coordinates of points A_1 and B_1 the A_1' and B_1'
   coordinates are computed:

$$\frac{A_1'O_1'}{A_1O_1} = \frac{B_1'O_1'}{B_1O_1} = m_a \tag{5.2}$$

where m_a is the scale factor or the coefficient of contour spread-out (a
function of the contour area); O_1 is the centre of gravity of the contour.

   For the raster image to be constructed it remains to connect these points
(A_1' and B_1') by a line with preassigned thickness and colour parameters. The
pairs of extreme points are taken with the frequency corresponding to the
accepted intervals between the raster axial lines.

3. Coordinates A_1' and B_1' are corrected so that the raster lines of neighbouring
   contours should not be superimposed in the case of the overlapping of
   contours when they are spread out.

Realization of the above-mentioned steps of the algorithm requires solving two
intricate problems:

1. The extreme points of contours (A_1B_1, A_2B_2, etc.) have to be identified.

The machine cannot distinguish segment A_1B_1 from segment A_4B_4, since it 'sees' not the whole map but only one processed line. From the input data it is only possible to determine whether a point belongs to the contour (code = 1) or to the background (code = 0). The same is true for segments A_2B_2 and A_3B_3. It is thus necessary to set the rule for the identification of extreme points.

2. From every contour point (say, one of the extreme points) we have to find the previously determined value for the coordinates of the centre of gravity and the respective scale factor, m_a, that is we must be able to assign each point to a certain contour.

Both these problems are solved using the procedure of 'painting' the contours.

To 'paint' a point means to replace its former code (0 or 1) with a certain symbol, designated as its 'colour'. This implies that if the previous point could occupy one bit, now it will have to be given several bits (in this case, seven). A line will consist of codes of the (0101101), (0110001), etc., types. The number of possible 'colours' in this algorithm was taken to equal $2^7 = 128$, which, as experiments have shown, is quite sufficient.

Code (0000000) is assigned to the points that do not belong to the contours proper. Code (1111111) is an auxiliary one and is not used for 'painting'. To 'paint' a contour is to 'paint' each one of its points and every point inside it. All the points of a contour are 'painted' identically. If two contours do not have common (i.e. lines containing the points of one and the other), then they can be 'painted' identically. This is done to economize on 'colours', as it is inexpedient to limit the number of contours on a map to 128. The procedure of 'painting' implies the process of the automatic recoding of points (in computer memory), and the word 'colour' is understood as a binary code, assigned to individual parts of a contour in the course of 'painting'.

Now the solution of the first problem, i.e. the rule for identifying the extreme points, is obvious. The solution of the second problem—how to ascertain that a point belongs to a concrete contour—is determined by its 'colour' and its y-coordinate. Thus, if the table of the parameters of contours, to which the coordinates of the centre of gravity and the values of m_z belong, includes also the 'colour' and the values of y_{max} and y_{min}, the necessary parameters for any two extreme points are then determined by exhaustive search until the following conditions are met:

the 'colour' = the 'colour' of the checked point;
$y_{min} < y$ (of the checked point) $< y_{max}$.

The general principle of contour 'painting' has just been examined, and now we shall describe this procedure in greater detail. It consists of three consecutive stages: (a) preliminary 'painting'; (b) formation of the control massif; (c) final 'painting'.

Preliminary 'painting'

The map is scanned from top to bottom line by line, and each newly appearing extension of a contour is 'painted' with a new 'colour'. If two different 'colours' are encountered in one line (Fig. 5.7), the 'colour' of the one on the right (in this case 'colour' A) is discontinued and after that the contour is 'painted' with the 'colour' of the left extension ('colour' D). 'Colour' A can be used again in other places. The fact of the jointing of two 'colours' is fixed in a special record.

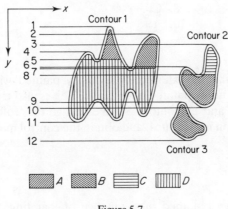

Figure 5.7

Thus, the results of the program run are: the 'painted' map; the table of 'colour' transitions. For Fig. 5.7 the record is as follows:

1. A	7. $B\emptyset \rightarrow D$
2. B	8. $C\emptyset \rightarrow A$
3. C	9. B
4. D	10. $A\emptyset$
5. $A\emptyset \rightarrow D$	11. $D\emptyset$
6. A	12. $B\emptyset$

The first four notations denote the initial points of those parts of the contours that are 'painted' with 'colours' A, B, C, D; the fifth denotes the end of 'colour' ($\emptyset$—the end of a 'colour') with transition into D, the sixth—the beginning of A (now in another region); the seventh—transition of B into D, with B discontinued; the eighth—C is discontinued and is followed by A; the ninth—beginning of B; from the tenth to the twelfth—the ends of 'colours' A, D, B.

The records thus contain all the data on the sequence of the beginning, the end, and the transition into each other of areas that should be 'painted' with one 'colour'.

Formation of the control massif

The control massif is the sequence of 'colours' with which newly appearing contour extensions have to be 'painted', so that every contour should later be 'painted' with only one 'colour'.

The records consist of groups belonging to one contour. In the present example these will be the following groups:

1. A	3. C	9. B
2. B	6. A	12. $B\emptyset$
4. D	8. $C\emptyset \to A$	
5. $A\emptyset \to D$	10. $A\emptyset$	
7. $B\emptyset \to D$		
11. $D\emptyset$		

It is therefore clear that the extensions previously 'painted' with 'colours' A, B, D (contour 1) have to be 'painted' identically. This is also true for 'colours' C and A (contour 2), and B (contour 3). We change the 'colours', taking into consideration division into groups, and obtain the control massif:

1. A	4. A
2. A	5. B
3. B	6. C

In this way we have obtained the sequence of 'painting' newly appearing extensions of contours. The final 'painting' will produce contour 1 'painted' with 'colour' A, contour 2 with 'colour' B, and contour 3 with 'colour' C.

Final 'painting'

The contours are 'painted' according to the sequence determined by the control massif, and the area of each contour is simultaneously computed from the number of its points. The coordinates of the centre of gravity are calculated from formulas:

$$x_T = \frac{\Sigma x_i}{S}, \quad y_T = \frac{\Sigma y_i}{S}$$

where x_i and y_i are the coordinates of the ith point and S is the area of the contour. Simultaneously calculated are the maximum and minimum coordinates of the contours, i.e. the least circumscribed rectangle with the sides parallel to the coordinate axes, for each contour.

The 'painting' proper is accomplished by successively scanning the map from top to bottom along the lines.

In composing a raster image it is necessary to prevent the superimposition of raster lines on each other. For this purpose, after spreading out a circumscribed rectangle m_a times, every pair of the contours is examined. If it so happens that these rectangles are superimposed on each other, a correction is made by shifting the raster lines upwards or downwards.

The generalization process requires special consideration. This is a multilevel process and can be fully automated. Owing to the multiscale representation of objects (contours) the selection process becomes unnecessary as all contours are shown. The only exception can be the small-scale mapping of territories with a high density (concentration) of contours. In this case the smallest contours are automatically selected and excluded.

The multilevel nature of the generalization process is a consequence of multiscale representation. There exist as many levels of generalization as there are scale factors. The degree of contour boundary generalization depends not only upon the scale (as it is usually understood) but also on the density of the lines filling a raster. It is evident that the denser the lines, the less the degree of generalization, and vice-versa. The degree of boundary generalization is thus a function of scale m_i and raster line density v_i, $g = F(m_i, v_i)$.

Figure 5.8 shows one and the same contour with the same scale factor but with a different areal weight, associated with a different density of raster lines. The contour in Fig. 5.8a is obviously subjected to greater generalization than the one in Fig. 5.8b, since after the readout and restoration of the contour boundaries by means of approximation by the end-points of the rarefied raster lines, the boundary will have a smoother outline than in the case of approximating by the end-points of the dense raster.

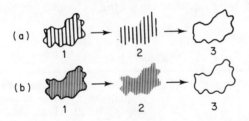

Figure 5.8

Taking into account the fact that the contours represented to scale are subjected to the highest degree of generalization, it is recommended the contour boundaries are shown. This technique improves the visual clarity of the image, as the contours represented to scale are more clearly distinguished from the contours plotted in other scales.

Conclusions

1. The technique shown makes possible a fully automatic preparation of the maps of areal objects, beginning with the readout and ending with engraving the smooth-delineation map.

2. The optimal solution in choosing the areal weights (scale factors) is searched for automatically in the course of map compilation.
3. The technique results in minimum data losses.
4. Apart from the direct problem—map compilation proper—the proposed technique allows one to solve other problems of independent significance:
   (a) determining the contour areas;
   (b) selecting the contours by the values of areas with the contours singled out within any preassigned gradation;
   (c) counting the contours, etc.

5.4.1. Comparison of the authenticity of areal objects representation on a normalized and a traditional map (as exemplified by forests)

The normalized map has been composed by the combined method (see Chapter 2) at a scale of $1:1,000,000$ from the maps of the $1:100,000$ scale. Three weights and, correspondingly, three scale factors (m_0, m_1, m_2) were taken on it. On the contours with the m_1 factor, shown by the lines of 0.4 mm thickness, 1 mm of line length accounts for 20 ha of forest area, i.e. $p_1 = 20$ ha, and on the contours with m_2 factor, shown by the lines of 0.1 mm thickness, 4 ha, i.e $p_2 = 4$ ha (Fig. 5.9 (a) multiscale and combined contours in the rasters (b) the boundaries of the contours plotted to the scale of the map and of the combined contours, (c) a portion of the traditional map).

The weights and the intervals between the raster axial lines within the generalized contours were calculated using the technique described in Chapter 2.

Table 5.1

	Initial map, scale $1:100,000$	Maps of $1:1,000,000$	
		Traditional	Normalized
Total area of forest contours represented on the map (km²)	280.64	363.36	288.57
No. of contours represented on the map:			
(a) total no. of contours	192	38	150
(b) contours combined into one		56	42
(c) contours not shown on the map		98	—
Sum of true errors in determining the area of each contour		+83.36	+7.93
Mean square error of contour area determination		6.27	0.05
Distortion of the shape of contours:			
(a) no. of undistorted contours		3	130
(b) no. of partially distorted contours		63	62
(c) no. of completely distorted contours		124	—

(a)   (b)

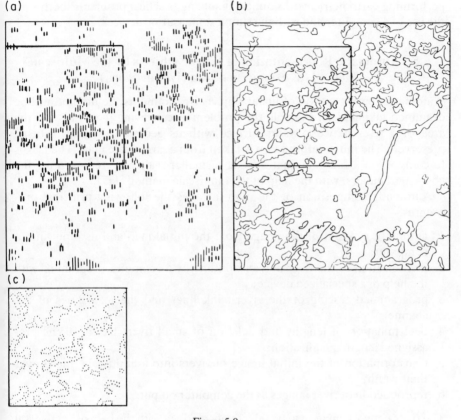

(c)

Figure 5.9

The results of comparative analysis of the normalized portion of a map, scale 1 : 1,000,000, with the traditional map of the same scale are shown in Table 5.1. Indicated in the table as partially distorted contours are those whose centre is preserved and the deviation from the true location of the boundaries does not exceed 3 mm. All the combined contours are also included in this group. The group of completely distorted contours accounts for the contours not shown on the map, as well as those whose centre of gravity is displaced by more than 3 mm from their true location. The contours whose boundaries were distorted by not more than 0.5 mm as a result of generalization with their overall configuration remaining intact are classified as undistorted. The centres of gravity of these contours exactly correspond to their true positions. This group includes all the small contours with the m_1 and m_2 scale factors, as their representation meets the requirements set.

The results of this analysis indisputably prove the advantage of the proposed method over the traditional ones in terms of the information content, accuracy, and authenticity of data representation, which is very important primarily in performing cartometric works aimed at solving practical problems, both with the use of a computer and with the simplest measuring devices.

5.5. Automatic Representation of Linear Objects by the Multiscale Method (as exemplified by a river network)

Automatic representation of linear objects by the multiscale method will be examined taking as an example the problem of conveying the width and the area of a river network water surface with its geometric similarity being preserved. The initial information is taken from a large-scale map. The task is to compose in a fully automatic mode a smaller-scale map representing the above parameters, with the help of the multiscale method [117].

Automatic composition of the map includes the following principal procedures:

1. automatic readout of the image from the initial map and its input into a computer;
2. identification of the river network outline by a computer program or with the help of a specialized device;
3. programmed tracing of the river-bank lines and the axial lines of the channels;
4. determination of lengths and deletion of small rivers based on the pre-assigned length qualification;
5. transformation of the initial image of rivers into a multiscale image over their width;
6. reproduction of river images at the computer output.

The river network image can be selectively read out with the help of a universal three-channel reading device performing the function of colour recognition (together with the function of analogue–digital conversion). It is quite obvious that the basic characteristic for the identification of a river network is colour (identification of symbols by their colour characteristics was considered in Chapter 4). True, selection of images by colour only does not altogether free the river network of such extraneous images as the designations of river depths, water edge elevations, lakes, and other symbols shown on the map with the same colour. Other additional parameters are therefore necessary for reliable identification to be achieved. Among the most characteristic parameters for river identification are: first the ratio between the length of the river-bank line and the water surface area, which is *a priori* higher than for lakes; then the parameter of sinuosity which permits one to distinguish rivers from canals, and the absence of breaks and closures—from letters. If the river network on the initial map is shown with a colour distinguishing it from all the other images, selective automatic readout is essentially simplified. The third and fourth procedures were dealt with in Chapter 4. As to selection of rivers by the preassigned length qualification, it is easily performed by a computer. It is very important in this case to find the means of representing the rivers falling under the preassigned criterion. Complete exclusion of the data on the number and spatial distribution of rivers, even of small length, diminishes the reliability of the map and makes it unsuitable for use in precise investigations.

The following principle of representation is proposed. Rivers falling under the preassigned selection criterion are shown at the points of their confluence with other rivers (shown on the map) as a short stroke. To do so it is expedient to subdivide the rivers into two or three gradations over their length and width. The smallest rivers are shown by a thin stroke and the larger ones by a thickened stroke.

Computer realization of this idea is reduced to a simple connection of several extreme points of the channel axis, lying at the mouth of the deleted river. As to the differentiation of rivers by their length, after determining the lengths of all the rivers it will again be only necessary to distribute them according to the preset gradations and then to reproduce them in the form of strokes of a certain thickness (in accordance with the gradation) and a constant length.

The remaining procedures will now be described in greater detail. As a result of the operations of tracing the bank lines of the rivers and identifying the points belonging to the median line of the channel, we shall obtain all the necessary points for the multiscale transformation of river images by their width. The procedure of multiscale transformation includes four basic operations:

1. determination of the channel width for river images from the median line and bank points;
2. classification of the river channels by this width according to the preassigned scale of gradations;
3. generalization of meanders;
4. determination of the new points of banks from the value of the preassigned scale factor and the obtained river width value.

The river width is determined along the normals to the median line. For transformation it is necessary to determine the river width at every point of the median line sample.

Let the ith point of the median line sample have coordinates $x_{0,i}$, $y_{0,i}$ (Fig. 5.10). What coordinates can point $(i + 1)$ have? If the coordinates of

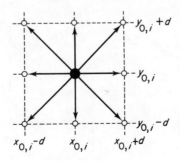

Figure 5.10

point $(i - 1)$ are disregarded there exist eight different possibilities for the location of point $(i + 1)$.

Let us analyse one of these eight possibilities. Let point i have the coordinates of $x_{0,i}$, $y_{0,i}$, and point $(i + 1)$ have the coordinates of $x_{0,i-d}$, $y_{0,i-d}$, which corresponds to the situation shown in Fig. 5.11a.

In this case point r with coordinates $x_{0,i+1}$, $y_{0,i}$ is examined (Fig. 5.11b). If this point coincides with a certain point of the river-bank sample, the value for the river width, c_i, is taken equal to twice the length of the perpendicular, r_i, dropped from this point to the segment connecting the ith and $(i + 1)$th points,

$$r_i = \sqrt{\left(\frac{x_{0,i} + x_{0,i+1}}{2} - x_r\right)^2 + \left(\frac{y_{0,i} + y_{0,i+1}}{2} - y_r\right)^2} \tag{5.3}$$

and $c_i = 2r_i$. If, however, this point coincides with none of the points in the river-bank sample, we examine the point with coordinates $x_{0,i+1}$, $y_{0,i} + d$ (Fig. 5.11c). If it belongs to the bank, we check the point with coordinates $x_{0,i+1} - d$, $y_{0,i}$ (Fig. 5.11d). If it is a point from the set we take the river width equal to twice the length of the perpendicular lying between the segments connecting points 1 and 2, i and $(i + 1)$ (Fig. 5.11e). If even one of the last two conditions is not met, we check the point with coordinates $x_{0,i+1} + d$, $y_{0,i} + d$, and so on.

Classification of river channels by their width consists in separating the rivers or channel sections into groups with respect to the preassigned intervals of river width values, following the accepted scale of gradations: $C_1, \ldots, C_n; C_n, \ldots,$ $C_{2n}; C_{2n}, \ldots, C_{3n}; C_{mn}, \ldots, C_{(m+1)n}$.

As a result of the preceding operation we shall obtain river width values at each point of the median line. Since a river channel sometimes has noticeable width variations, in order to avoid unnecessary diminution of gradations it is recommended that the mean width of the channel be determined from several successive values of its width

$$\bar{C}_p = \frac{1}{n_p'} \sum^{n_p'} c_i$$

where p is the number of the segment ($p = 1, 2, \ldots, N$). In other words, the rivers are divided into sections of n_p' points in each. For each of them, from the

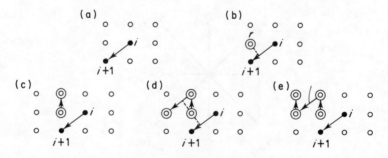

Figure 5.11

width values obtained in the preceding operation, we find the mean width values, $\overline{C}_p$, which are then grouped by the gradations with the ends of the sections being fixed. The length of the sections, or the number of successive channel width values, from which $\overline{C}_p$ is found, is determined proceeding from the scale of the newly composed map, the purpose of the map, as well as the region of mapping.

When generalizing the river meanders in going to smaller-scale maps, some meanders of a linear object will naturally have to remain unshown because of the limitations imposed by the image representation facilities (primarily by the printing equipment) and the perception of the human visual analyser. Such generalization is enforced and is not aimed at solving any concrete problem. For analytical maps, and for topographic maps as well, a detailed and accurate representation of information about the linear objects, with minimal generalization of meanders, is very important.

Before generalizing the meanders it is necessary to establish certain formal criteria, based on which some meanders should be smoothed out and the others preserved. Let us first define a meander. A meander is understood as a portion of unbroken curve (primarily representing a river image) limited by points of zero curvature, or inflection points. If all the zero curvature points are connected with straight lines the whole curve is approximated by a polygonal line. In this case, polygonal line $a_{i-1}, a_i, a_{i+1}, \ldots, a_{i+k}$ will divide the whole curve into individual meanders (Fig. 5.12a) of different length. Any meander can be characterized with sufficient comprehensiveness (for the given case of scale generalization) by its length ΔL, the distance, l_0, between the zero curvature points, and the minimum distance, $l_{\min}$, between adjacent meanders, M and F (Fig. 5.12b). To establish the criteria for meanders generalization let us take these three parameters. We shall adopt the following criteria:

1. if l_0 or $l_{\min} < \overline{C}_p K_q + \mu$, and $\Delta L < \frac{1}{2}\pi C'_{\min} + \theta_q$, the meander is excluded;

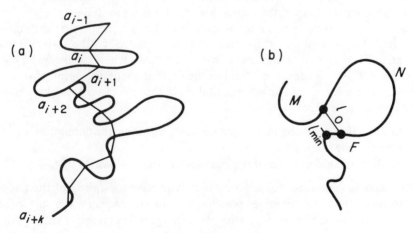

Figure 5.12

2. in all other cases the meanders are preserved. Here μ is the minimum permissible distance between lines (to the scale of the newly compiled map), it is determined taking into account the visual perception and the capacity of image reproduction equipment; θ_q is a quantity established depending on the purpose of the map, and on the significance and natural features of the mapped object; $C'_{min} = C_{min} \times K$, where C_{min} is the minimum value of river channel width on the initial map (Fig. 5.13).

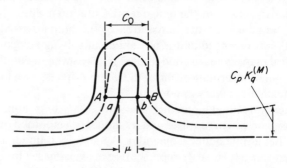

Figure 5.13

On the strength of

$$AB \approx Aa + ab + bB, \quad Aa \approx \tfrac{1}{2}\overline{C}_p \times K_q$$
$$a \approx \mu, \quad bB \approx \tfrac{1}{2}\overline{C}_p \times K_q, \quad AB = l_0$$

the meaning of the first criterion becomes obvious. The second and the third criteria are to a certain extent arbitrary, particularly the limitation for ΔL. In the first criterion, expression $\tfrac{1}{2}\pi C'_{min}$ is the minimum (threshold) length of the meander represented on the map. Note that for wide rivers, represented to the scale of the compiled map, meanders can be generalized not along the median line but along the lines of the banks, for each bank separately.

In the case when a meander is excluded, i.e. when the first criterion is satisfied, the median line of the river channel is plotted by the zero curvature points. The second criterion allows the case when a meander with $l < C_p \times K + \mu$ is preserved, but this means that contiguity or intersection of the contour lines of the banks is inevitable. In this case, two possible solutions can be recommended:

1. to shift the river channels by a distance ensuring the minimum gap, μ, between the banks;
2. to show the median line of the channel by dots or a dashed line.

The second variant makes a geometrically precise indication of the channel position on the meander image possible, but the accuracy of water surface area representation and the visual clarity of the image will decrease. The first variant yields the opposite result and requires a more complex processing. Moreover,

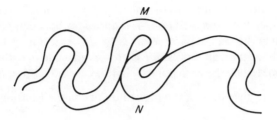

Figure 5.14

in a number of cases computer implementation of the first variant proves to be very difficult. It is especially difficult to shift the channels for such meanders, as shown in Fig. 5.14.

Two meanders following one another, M and N, merge in narrow places. It is very difficult to shift the channels for meanders M and N following the first variant as the shift of one of them aggravates the position of the other; besides, this may lead to a serious alteration of the true positions of the channels. For medium-size and large rivers, i.e. those whose width permits showing the position of the middle of the channel with a dashed line, it is expedient to use the second variant. Only for small rivers, where it is impossible to show the median line, a shift is expedient, but only for individual large meanders, with the small ones being deleted.

Let us now consider three cases of the relative position of meanders when a shift of the channels becomes necessary:

1. meanders M and F, adjacent to the meander in question, N, are not preserved in accordance with the adopted criteria (i.e. they are approximated by zero points), and the distance between zero points A_0 and B_0 of meander N is less than the sum of parameters $\overline{C}_p \times K + \mu$ (Fig. 5.15a);

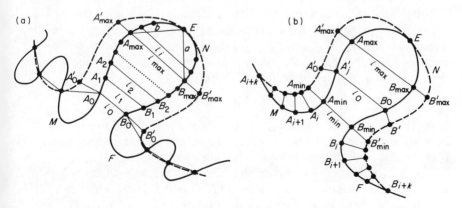

Figure 5.15

2. meanders M and F, adjacent to the meander in question, N, are preserved, and the distance between zero points A_0 and B_0 or between points $A_{\min}$ and $B_{\min}$ is less than the sum of parameters $\overline{C}_p \times K_q + \mu$ (Fig. 5.15b);

3. one of the adjacent meanders M or F is preserved while the other is not (i.e., it is approximated by the zero points), in this case $l_0 < \overline{C}_p \times K_q + \mu$.

Let us examine the algorithm of the procedure of channel shifting for the three above cases.

1. For the first case, as well as for the other two, it is recommended that the channels be shifted at the median line points, and that the shift be performed in two opposite directions, perpendicular to the median line. The magnitude of shift σ_0 is first determined for the points of zero curvature, A_0 and B_0, which is calculated from formula:

$$\sigma_0 = \tfrac{1}{2}(\overline{C}_p \times K_q + \mu - l_0) \tag{5.4}$$

Both the zero curvature points are shifted along the perpendicular to a distance corresponding to the obtained σ_0 value. As a result, they will occupy positions A_0' and B_0' (see Fig. 5.15a). Then points $A_{\max}$ and $B_{\max}$ are found, with a maximum distance between them of $l_{\max}$. These points are searched for by successive exhaustion of pairs of points $A_1B_1, A_2B_2, \ldots, A_iB_i$ and by determining the distances between them, $l_1, l_2, \ldots, l_i$; the $l_2 - l_1, l_3 - l_2, \ldots, l_i - l_{i-1}$ differences are also determined. Once a situation arises where $e_i - e_{i-1} < 0$, the previous pair of points is taken as representing the maximum distance, that is $A_{i-1} = A_{\max}, B_{i-1} = B_{\max}$. The found $A_{\max}$ and $B_{\max}$ points are shifted along the perpendicular to a distance of σ_0 to occupy the positions of $A_{\max}$ and $B_{\max}$, respectively. After that the extremum, E, of meander N is found. Point E is determined searching for a triangle most closely approximating the arc limited by points $A_{\max}$ and $B_{\max}$. The vertex of this triangle is taken as the point of extremum. The base of the triangle coincides with the $A_{\max}$ and $B_{\max}$ side. The difference, $A_{\max} - B_{\max} - (b + a)$, must be minimal in comparison with all the possible triangles.

The procedure of constructing such a triangle begins with searching for an approximate extremum, E', which is found as the point having the medium number in between $A_{\max}$ and $B_{\max}$. We shall designate the number of the $A_{\max}$ point as i and that of the $B_{\max}$ point as k, then the number of the E' point, designated as q, will be obtained from expression.

$$q = \tfrac{1}{2}(i + k) \tag{5.5}$$

From the coordinates of points $A_{\max}$, $B_{\max}$, and E we find the sum of two sides; $\Sigma_q = a_q + b_q$ (see Fig. 5.15a). Then, for the neighbouring points with numbers $(q + 1)$ and $(q - 1)$, we shall find the sum of sides $\Sigma_{q+1} = a_{q+1} + b_{q+1}$ and $\Sigma_{q-1} = a_{q-1} + b_{q-1}$ and compare the obtained values with $\Sigma_q > \Sigma_{q+1}$ and $\Sigma_q > \Sigma_{q-1}$, then point E' is the extremum E. Otherwise the search continues in the direction of the point for which the sum of the sides was larger than Σ_q. Similarly we continue to calculate the sums of the sides successively for points in the order of their numbers, beginning with the point for which the sum of the sides was larger than Σ_q. Each time we take the difference of the sums Δ, i.e. the sum of the sides adjoining the subsequent point minus the sum of the sides adjoining the preceding point. When the situation arises where $\Delta < 0$, the preceding point is taken for the extremum. The extremum, E_i, can be left at its place or shifted to a distance of σ_0 outwards along the normal. This depends on the aims of generalization.

The characteristic points, A_0', A_{max}', E, B_{max}', B_0', of the meander $A_0 B_0$, obtained in this manner, are then used, together with the zero curvature points of the adjacent meanders, to construct smooth curves representing the generalized image of the meander. Construction of these curves is described at the end of the section (in Fig. 5.15 the generalized median line of the channel is shown by a dashed line).

2. In the second case we begin with searching for the minimum distance, l_{min}, between meanders M and F, i.e. with determining the distance between the most closely lying points A_{min}, B_{min}, if such points exist (see Fig. 5.15b). Starting from the points of zero curvature, the distances between successive points of meanders M and F: $l_1, l_2, l_3, \ldots$ are determined. If $l_2 < l_1$, or $l_3 < l_2$, the minimum exists; otherwise the search is discontinued. If l_{min} is established to exist, it is searched for. As soon as the situation arises where $l_i > l_{i-1}$, the search is discontinued. The points corresponding to A_{min} and B_{min} are A_{i-1} and B_{i-1}. The magnitude of shift, σ_{max}, is then determined for points A_{min}, B_{min} from the formula:

$$\sigma_{max} = \tfrac{1}{2}(C_p \times K_q + \mu - l_{min}) \tag{5.6}$$

As in the first case, points $A_0, B_0, A_{max}, B_{max}$ are shifted by the value of σ_{max} and the extremum E is found. Then the values of σ_i, $\sigma_{i+1}, \ldots, \sigma_{i+k}$ are found for points A_i and B_i, A_{i+1} and $B_{i+1}, \ldots, A_{i+k}$ and B_{i+k}, respectively, from the formula:

$$\sigma_i = \tfrac{1}{4} 2(\overline{C}_p \times K_q + \mu) - (l_{i-1} + l_i) \tag{5.7}$$

The points are shifted until $\sigma_{i+k} \leqslant d$, where d is the discretization step. In the case when l_{min} does not exist, a similar procedure is performed beginning from points A_0 and B_0. Note that in general it is unnecessary to determine σ_i for all the median line points of the channel of the initial map. If the compiled map is diminished m times the points should also be selected with an interval corresponding to m points, i.e. A_i and B_i, A_{i+m} and B_{i+m}, A_{i+2m}, and B_{i+2m}, etc.

3. Finally, the third case, when one of the meanders M or F is preserved and the other one is not. Here, as in the first case, the same procedure is performed with meander N and then, beginning from points $A_0 B_0$, the procedure is the same as in the second case, but applied to the meander being preserved.

The new points of the banks are determined, in accordance with the scale of gradations and the classification of channels, at the median line points preserved as a result of the generalization of meanders. For this purpose the values of C_i, obtained in the first operation, are multiplied by the scale factor in accordance with the classification (grouping) of river channels by their gradations, $C_i' = C_i \times K_q$. The coordinates of the new river bank are the ends of perpendiculars r_j' to both banks constituting the extensions of the perpendiculars r_j obtained in the first operation.

When dealing with rivers having a complicated bank line (which is characteristic of large rivers), it is necessary to determine the perpendiculars to each river bank separately. For small rivers a perpendicular to one bank is sufficient, and it is then extended to the other bank with the same length value. It should be noted that, because of the meandering of rivers, the points of the banks are distributed with a varying interval between them. A bank with a larger curvature has a higher density of points.

Described below is a practically realized program of fully automatic generalization. The operation of the program is examined after the river-bank lines have been traced, beginning with the determination of channel median line points from the following formulae:

$$\left. \begin{aligned} x_i &= (x_{bi} + x_{b1i})/2 \\ y_i &= (y_{bi} + y_{b1i})/2 \end{aligned} \right\} \tag{5.8}$$

where (x_{bi}, y_{bi}) are the coordinates of the ith point of the right bank; (x_{b1i}, y_{b1i}) are the coordinates of the ith point of the left bank.

In the next operation we determine the channel width for the initial river image. Channel width at the ith point is taken equal to the doubled length of the perpendicular lying between the ith and $(i + 1)$th points of the median line and the corresponding points on the bank. In this operation two segments, given by the following equations, are treated:

$$y(x_{bi+1} - x_{bi}) - x(y_{bi+1} - y_{bi}) + x_{bi}(y_{bi+1} - y_{bi}) - (y_{bi+1} - x_{bi}) = 0 \tag{5.9}$$

$$y = \frac{x_i - x_{i-1}}{y_{i+1} - y_i} x + \frac{y_i + y_{i+1}}{2} - \frac{x_i - x_{i+1}}{2} = 0 \tag{5.10}$$

The first one is the equation of the straight line passing through two adjacent points on the bank (x_{bi}, y_{bi}), (x_{bi+1}, y_{bi+1}). The second is the equation of the straight line perpendicular to the segment connecting two adjacent points on the river channel median line, passing through the middle of this segment.

Solving these equations jointly we find the intersection point of these straight lines:

$$\begin{aligned} x_2 &= [(y_{i+1} - y_i)(y_{bi+1} - y_{bi})x_{bi} + (x_{bi+1} - x_{bi})(y_{i+1} - y_i)(y_i + y_{i+1})/2 + \\ &\quad + (x_{i+1} - x_i)(x_{bi+1} - x_{bi})(x_i + x_{i+1})/2 - (x_{bi+1} - x_{bi})(y_{i+1} - y_i)y_{bi}] \div \\ &\quad \div [(y_{bi+1} - y_{bi})(y_{i+1} - y_i) + x_{bi+1} - x_{bi})(x_{i+1} - x_i)] \\ y_2 &= y_{bi} + (y_{bi+1} - y_{bi})(x_2 - x_{bi})/(x_{bi+1} - x_{bi}) \end{aligned}$$

$$\tag{5.11}$$

The width of the river image at the ith point is taken equal to the doubled distance between the point obtained and the central point of the segment of the channel median line:

$$C_i = 2_{ri} = [(x_2 - x_c)^2 - (y_2 - y_c)^2]^{-1/2} \qquad (5.12)$$

where $x_c = (x_i + x_{i+1})/2$, $y_c = (y_i + y_{i+1})/2$.

In the course of determining the width, the maximum width of the image is also found. If this width proves to be below the qualification value preassigned in the program, then after the image has been generalized the coordinates of the new points of the banks are not computed and the message 'BANK LINES MERGE' is printed. In this case the derivative image is constructed on new points of the median line of the generalized channel.

The next stage is the generalization of the meanders of the river image. Every meander is confined between points of zero curvature. Let us consider the principle of determining the points of zero curvature.

Let (x_i, y_i), (x_{i+1}, y_{i+1}), (x_{i+2}, y_{i+2}), (x_{i+3}, y_{i+3}) be the successive four points on the median line of the initial river channel image.

A straight line is drawn through two adjacent points (x_{i+1}, y_{i+1}) and (x_{i+2}, y_{i+2}), and its equation has the following form:

$$B(x, y) = (y_{i+2} - y_{i+1})(x - x_{i+1}) - (x_{i+2} - x_{i+1})(y - y_{i+1}) = 0 \qquad (5.13)$$

Points x and y are substituted by the first and the last coordinates of the four, i.e. (x_i, y_i), (x_{i+3}, y_{i+3}). If this results in $B(x, y)$ assuming values of different sign, there exists a point of inflection between these coordinates, if the signs are the same, there is no inflection. In the latter case we pass on to the next four points and check them for inflection. If the checking has confirmed the presence of a point of inflection, it is necessary to identify one of the two available points: either (x_{i+1}, y_{i+1}) or (x_{i+2}, y_{i+2}). For this purpose four points, (x_{i+1}, y_{i+1}), (x_{i+2}, y_{i+2}), (x_{i+3}, y_{i+3}), (x_{i+4}, y_{i+4}), are considered and checked. If there is no inflection between (x_{i+1}, y_{i+1}) and (x_{i+4}, y_{i+4}), it means that the sought point of inflection is (x_{i+1}, y_{i+1}), otherwise it is (x_{i+2}, y_{i+2}).

Having found two adjacent points of inflection and, consequently, the meander confined between them, we check whether it meets the generalization criteria:

1. if $l_0 < C_i K_q + \mu$ the meander is excluded;
2. otherwise the meander is preserved,

l_0 is the distance between the points of inflection;
C_i is the channel width at the ith point;
K_q is the scale factor;
μ is the minimum permissible space between the lines (to the scale of the composed map).

In the case when condition 1 is fulfilled the median line of the river is plotted through the zero curvature points.

The stage of width transformation ends there, and length transformation begins. Here it is necessary to reduce the distance between points K_q times. To do this we multiply the coordinates of the median line points by the scale factor, taking into account the image shift resulting from it.

The last operation of the automatic construction of river channel image from the available large-scale image is to determine new points of the banks (provided the bank lines do not merge). They are determined at the preserved points of the median line. The coordinates of the new banks are the ends of perpendiculars r'_j ($r'_j = r_j \times K_q$) to both banks, constituting the extensions of perpendiculars r_j obtained when the width of the channel image was computed.

Solving again a set of equations:

$$\left. \begin{aligned} y &= \frac{x_i - x_{i+1}}{y_{i+1} - y_i} x + \frac{y_i + y_{i+1}}{2} \frac{x_i - x_{i+1}}{y_{i+1} + y_i} \frac{x_i + x_{i+1}}{2} \\ r_i'^2 &= [x - (x_i + x_{i+1})/2]^2 + [y - (y_i + y_{i+1})/2]^2 \end{aligned} \right\} \quad (5.14)$$

we find the analytical expression for the coordinates of new points of the banks:

$$\left. \begin{aligned} y'_i &= (y_i + y_{i+1})/2 + r'_i/[((y_{i+1} - y_i)/(x_i - x_{i+1}))^2 + 1]^{-1/2} \\ x'_i &= (x_i + x_{i+1})/2 + [(y_{i+1} - y_i)/(x_i - x_{i+1})]/(y'_i - (y_{i+1} + y_i)/2) \end{aligned} \right\} \quad (5.15)$$

and

$$\left. \begin{aligned} y'_{1i} &= (y_i + y_{i+1})/2 - r'_i[((y_{i+1} - y_i)/(x_i - x_{i+1}))^2 + 1]^{-1/2} \\ x'_{1i} &= (x_i + x_{i+1})/2 + [(y'_{1i} - (y_{i+1} + y_i)]/2 \times [(y_{i+1} - y_i)/(x_i - x_{i+1})] \end{aligned} \right\} \quad (5.16)$$

The results of the program are represented by a table of the coordinates of new bank points, a group of messages concerning the excluded river channel meanders, and by the generalized river image.

5.5.1. Analysis of the reliability of linear objects representation on a normalized map as compared to a traditional map (as exemplified by a river network)

The normalized map was composed (by the method described in Chapter 2) using several scales: for the rivers represented by two lines the scale factor was K_1, for the rivers with a width between 10 and 60 metres it was K_2, and for the rivers with a width less than 10 metres, unrepresentable to the scale of the compiled map, it was K_3.

A conventional symbol of a river unrepresentable by two lines is a dashed line, the lengths of dashes not being constant. This results from the special way of constructing a normalized map. On the initial cartographic material, where the density and sinuosity of the river network are represented in detail, the lengths of rivers were measured with a constant step from the mouth to the source, or from one edge of the map to the other when the source was outside the map. In our case such measurements were made on a topographic map of 1 : 10,000 scale with a constant step of 200 m. This was followed by cartographic generalization performed in accordance with the general rules for topographic maps, as a result of which the number of meanders and the river lengths changed. At the sites where the number of meanders and, consequently, the river length is drastically reduced the dashes are considerably shorter than at the less generalized sites. Visual calculation of the length of a river on such a map is quite simple: the number of dashes is multiplied by the constant step of 200 m.

Intermittent rivers are assigned the conventional symbol of a dotted line. The same principle is valid here: all points on the initial map are located at strictly equal distances from each other, corresponding to the constant step of river measurement; on the composed map the intervals between the points vary depending on the generalization performed. The length of the intermittent part of a river is in this case represented by the product of the number of the points by the constant step (100 m). For large rivers, whose width is represented to the scale of the composed map, measurements are made along the central points of the channel. Projections of measurement points are shown by dots. When a river is very wide, the measurements are made along both banks, and the bank line is shown by a dashed line in accordance with the same rule.

This particular technique of composing a normalized map can be applied in plotting specialized thematic maps requiring a high accuracy of the represented characteristics of linear objects.

To estimate the accuracy of the technique in question a comparative analysis of the composed normalized map (scale 1 : 100,000) and a traditional topographic map of a larger scale (1 : 50,000) was performed. The characteristics of linear objects represented on the basic map of a 1 : 10,000 scale (Table 5.2) were taken as the reference characteristics.

Table 5.2

No. of rivers	Results of river length measurements in metres			Values of distortions caused by generalization	
	Initial $S : 10,000$	Normalized $S\, 1 : 100,000$	Traditional $M\, 1 : 50,000$	Normalized	Traditional
1	3690	3700	3700	+10	+10
2	3500	3500	3280	0	−220
3	2350	2350	2110	0	−240
4	2290	2300	2200	+10	−90
5	490	490	310	0	−180
6	920	925	1000	+5	+80
7	1200	1200	1150	0	−50
8	2400	2400	2320	0	−80

188

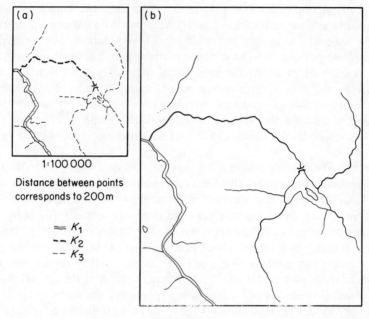

Figure 5.16

Figure 5.16 shows a portion of the map containing only the linear elements of specific map content (a = multiscale image, b = initial image).

It is seen from the table that while the normalized map has a much smaller scale (1:100,000) than the traditional map compared with it (1:50,000), the accuracy of measurements on the normalized map is markedly higher.

A normalized river network map allows one to make accurate calculations of river lengths and water surface areas, to judge the true frequency of meanders on a certain river, and to perform automatic generalization when composing smaller-scale maps. It also facilitates manual cartometric work.

CHAPTER 6
Certain Aspects of the Automatic Transformation and Interpretation of Aerospace Photographs and Maps

The principles of automatic processing of aerospace photographs and of maps, as graphical data carriers, have much in common. In studying the objects situated on the Earth's surface and their thematic mapping it is expedient, from considerations of economic and technical efficiency, to process aerospace photographs jointly with maps. For instance, to correct photographic images of the Earth's surface relief it is more effective to begin with composing a digital model based on topographic maps, using normalization methods, and then correct, with its help, the photograph on a computer. Integrated processing of maps and photographs is particularly effective in solving the problems of forecasting the development in natural processes, when the models used incorporate the parameters relating to long-term, essentially unchanging, information. At present there exist a large number of different types of maps showing the data characterized by very slow changes. Among this kind of data one can mention: the geometric presentation of relief on topographic maps, qualitative soil characteristics on pedological maps, groundwater characteristics on hydrogeological maps, and many others.

Methods of transforming the images of extensive objects, considered in this chapter, are equally applicable both to maps and aerospace photographs.

Specific features characterize the use of aerospace photographs and of maps in solving the problems of aerospace navigation. Using normalized maps in navigation appears to be of greatest interest.

6.1. The Use of Space Photographs and Maps for the Purposes of Space Navigation

6.1.1. Space navigation by remote sounding data

Problems connected with space navigation impose special requirements on the processing of space photographs. Photographic films are usually developed in terrestrial conditions, which limits to a large degree the promptness and economic efficiency of obtaining the required data. As to television equipment, it has enhanced power consumption, small capacity and noise immunity, and can function only within the line-of-sight distance between the space vehicle (SV) and the ground data receiving and processing posts (GDP). It would therefore be expedient to preprocess video data directly on board the SV with the object of 'compressing' them to a minimum of volume and reducing them to a form most convenient for on-board storage and subsequent remote transmission to the Earth. Data acquisition in three stages appears to be the most appropriate.

At the first stage time-lapse photography of the planet surface is performed from manned SV; each photograph is approximated by one or several contrast levels with the help of quadratic curves, and the invariants of the approximating curves are subsequently used as image characteristics. At this stage a limited file of sample images of the planet surface is collected, and these images, in the form of photographs and the corresponding sets of numerical values, are returned to the Earth in descent modules. As a result of the first stage a master file of the 'standard' images for the investigated planet will be produced together with the corresponding file of numerical values for the characteristics unambiguously describing these images.

At the second stage only automatic unmanned SV equipped with image scanning and coding systems are used. This stage consists of automatic frame-by-frame coding of the planet images and their coordinate fixation with respect to the trajectory of the automatic unmanned SV. Photographic recording of the planet images is not performed. They are scanned in real-time mode directly from the screen of an optical, infrared, or radiolocation display sight; and the image scanning, reading, and coding are performed, as at the first stage, with the help of a reading device of the flying-spot tube type. For each quantization point the on-board digital computer records and stores the voltage value (in coded form) proportional at these points to the contrast factor of the scanned image. The whole range of output values, from V_{min} to V_{max}, is subdivided into m equal sections assigned a certain contrast level. As a result a certain vth section of this range will have the corresponding voltages lying within the interval from

$$V_{min} + (v - 1) \frac{V_{max} - V_{min}}{m} \text{ to } V_{min} + v \frac{V_{max} - V_{min}}{m}$$

Location of the points corresponding to a certain contrast level is approximated by a quadratic curve in a planimetric Cartesian coordinate system. The curves for each contrast level are plotted as follows. Coordinates x_i and y_i ($i = 1, 2, \ldots, N$) for the points of the image with this contrast level are determined and recorded in the on-board digital computer storage. Each pair of x_i, y_i values is substituted into an equation of the form:

$$b'xy + c'y^2 + d'x + e'y + f = -x^2$$

obtained from the canonical equation of a quadratic curve,

$$ax^2 + 2bxy + cy^2 + 2dx + 2ey + f = 0$$

after dividing it by the factor at the first term and transposing the first term of the equation obtained into the right-hand side. As a result we get a set of N equations with five unknown quantities: b', c', d', e', f':

$$\left.\begin{array}{l} b'x_1y_1 + c'y_1^2 + d'x_1 + e'y_1 + f' = -x_1^2 \\ b'x_2y_2 + c'y_2^2 + d'x_2 + e'y_2 + f' = -x_2^2 \\ \cdots\cdots\cdots\cdots\cdots\cdots\cdots\cdots\cdots \\ b'x_{N-1}y_{N-1} + c'y_{N-1}^2 + d'x_{N-1} + e'y_{N-1} + f' = -x_{N-1}^2 \\ b'x_Ny_N + c'y_N^2 + d'x_N + e'y_N + f' = -x_N^2 \end{array}\right\} \quad (6.1)$$

Note that the system obtained is a set of linear equations. If we write out this set of equations in matrix form, then, using least-squares method, factors b', c', d', e', and f', can be obtained

$$\mathbf{V}^T\mathbf{PVA} = \mathbf{VPW} \tag{6.2}$$

where: $\mathbf{V}$ is the matrix of known quantities in the left-hand side of the set (6.1), determined by values x_i and y_i;

$\mathbf{V}^T$ is the transposed matrix;

$\mathbf{W}$ is the matrix consisting of the terms in the right-hand side of the set (6.1);

$\mathbf{A}$ is the matrix of estimates (of sought unknown quantities), determined by the least squares method;

$\mathbf{P}$ is the diagonal matrix of weights (in our case, the weights of x_i and y_i measurement errors).

The errors of measuring the values of x_i and y_i in the course of image readout by a cathode-ray tube and in subsequent electronic processing have numerous causes: scanning non-linearity, instability of time-base generators, quantization, etc. The magnitude of these errors, however, does not exceed one element of resolution, and the distortion of coordinates for the points of the identified image contour within such limits does not actually affect the values of the characteristics. Consequently, the weights of errors can be regarded as identical and equal to a unity, and the matrix $\mathbf{P}$—as a unit matrix. Then equality (6.2) will be written in the form

$$\mathbf{V}^T\mathbf{VA} = \mathbf{V}^T\mathbf{W} \tag{6.3}$$

The matrix of **A** estimates is determined unambiguously.
We thus have:

$$
\mathbf{V} = \begin{Vmatrix}
x_1 y_1 & y_1^2 & x_1 & y_1 & 1 \\
x_2 y_2 & y_2^2 & x_2 & y_2 & 1 \\
. & . & . & . & . \\
x_N y_N & y_N^2 & x_N & y_N & 1
\end{Vmatrix}
\tag{6.4}
$$

$$
\mathbf{V^T} = \begin{Vmatrix}
x_1 y_1 & x_2 y_2 & \cdot & x_N y_N \\
y_1^2 & y_2^2 & \cdot & y_N^2 \\
x_1 & x_2 & \cdot & y_N \\
y_1 & y_2 & \cdot & y_N \\
1 & 1 & \cdot & 1
\end{Vmatrix}
\tag{6.5}
$$

$$
\mathbf{W} = \begin{Vmatrix}
-x_1^2 \\
-x_2^2 \\
. \ . \ . \\
-x_N^2
\end{Vmatrix}
\tag{6.6}
$$

$$
\mathbf{A} = \begin{Vmatrix}
b' \\
c' \\
d' \\
e' \\
f'
\end{Vmatrix}
\tag{6.7}
$$

The matrix product has the form of

$$
\mathbf{V^T V} = \begin{Vmatrix}
[x^2 y^2] & [xy^3] & [x^2 y] & [xy^2] & [xy] \\
[xy^3] & [y^4] & [xy^2] & [y^3] & [y^2] \\
[x^2 y] & [xy^2] & [x^2] & [xy] & [x] \\
[xy^2] & [y^3] & [xy] & [y^2] & [y] \\
[xy] & [y^2] & [x] & [y] & [1]
\end{Vmatrix}
\tag{6.8}
$$

where $[x^2 y] = \Sigma_{i=1}^{N} x_i^2 y_i$; $[xy^3] = \Sigma_{i=1}^{N} x_i y_i^3$, etc. $|\mathbf{V^T V}|$ is the Gramian determinant of system (6.1). Since the rank of $|\mathbf{V^T V}| = N \neq 0$, the set of equations is solvable.

According to expression (6.3), taking into account (6.2), (6.4) and (6.8), we obtain a set of n equations with n unknown quantities:

$$
\begin{aligned}
[x^2 y^2]b' + [xy^3]c' + [x^2 y]d' + [xy^2]e' + [xy]f' &= -[x^3 y] \\
[xy^3]b' + [y^4]c' + [xy^2]d' + [y^3]e' + [y^2]f' &= -[x^2 y^2] \\
[x^2 y]b' + [xy^2]c' + [x^2]d' + [xy]e' + [x]f' &= -[x^3] \\
[xy^2]b' + [y^3]c' + [xy]d' + [y^2]e' + [y]f' &= -[x^2 y] \\
[xy]b' + [y^2]c' + [x]d' + [y]e' + [1]f' &= -[x^2]
\end{aligned}
\tag{6.9}
$$

Having solved this set with respect to b', c', d', e', f', and dividing the values of factors by two, we shall get factors

$$b'' = \frac{b'}{2}, \quad c'' = c', \quad d'' = \frac{d'}{2}, \quad e'' = \frac{e'}{2}, \quad f'' = f'$$

used, at $a'' = a' = 1$, to calculate invariants:

$$\Delta = \begin{vmatrix} a'' & b'' & d'' \\ b'' & c'' & e'' \\ d'' & e'' & f'' \end{vmatrix} ; \quad \gamma = \begin{vmatrix} a'' & b'' \\ b'' & c'' \end{vmatrix} ; \quad \text{and } S = a'' + c'' \tag{6.10}$$

where Δ and γ are known in analytical geometry as a large and a small discriminant, respectively.

These parameters characterize the image for a given fixed level of its contrast range. They do not depend on the position of the image in the reading device field of view; they are therefore not critical with respect to the direction of the SV movement relative to the planet surface and permit considerably lowering the requirements on the accuracy of SV attitude and stabilization control systems.

The values of Δ, γ, and S, together with the coordinate or time fixation data for each of the j images, are recorded in a coded form in the on-board storage and, in accordance with the program or by commands from the Earth, are transmitted through the telecommunication channel to the GDP.

The third stage is actualized directly at the ground data receiving and processing post. It involves successive comparison of the sets of invariants Δ_j, γ_j, S_j, obtained at the second stage, with the respective sets of invariants, Δ_{0j}, γ_{0j}, S_{0j}, for 'standard' images, obtained at the first stage. If the comparison shows that within the limits of permissible tolerances the equalities, $\Delta_j = \Delta_{0j}$, $\gamma_j = \gamma_{0j}$, $S_j = S_{0j}$, are valid, the decision is taken to identify the image (photograph) represented by the Δ_j, γ_j, S_j invariants with the reference image whose invariants are Δ_{0j}, γ_{0j}, S_{0j}. If $\Delta_j \neq \Delta_{0j}$, but $\gamma_j = \gamma_{0j}$ and $S_j = S_{0j}$, the images obtained at the second stage are reduced to the scale of the standard images, in accordance with expression

$$N = (\Delta_j/\Delta_{0j})^{1/2} \tag{6.11}$$

This is required only if SV orbital altitudes at the first and the second stage have proved to be different. It should be noted that the use of the above method of coding and remote data acquisition is restricted in the optical wavelength band by the necessity to scan and code video images at the second stage under the same lighting conditions on the planet surface as at the first stage. These restrictions do not exist when the data are obtained from radar images of the planet surface. In this case, however, certain difficulties arise in connection with converting the radar images into conventional video images obtained in the optical wavelength band.

To avoid data losses in the readout of images the resolving power of the reading system must correspond to that of the ground observation system.

There is no point in trying to attain the readout system (e.g. of the CRT type) resolution much higher than that on the terrain, since this not only gives no additional information, but also prolongs the reading time and introduces operational errors into identification algorithms, critically dependent on scale changes of the image.

In experiments, system (6.9) was solved by the Gaussian method without return running with the choice of the leading element in a column. As applied to system

$$
\left.\begin{aligned}
a_{11}x_1 + a_{12}x_2 + \cdots + a_{1n}x_n &= a_{1(n+1)} \\
a_{21}x_1 + a_{22}x_2 + \cdots + a_{2n}x_n &= a_{2(n+1)} \\
a_{n1}x_1 + a_{n2}x_2 + \cdots + a_{nn}x_n &= a_{n(n+1)}
\end{aligned}\right\}
\tag{6.12}
$$

this method is reduced to performing the following sequence of operations.

The first equation is divided by a_{11}, then successively multiplied by a_{i1} and subtracted from the ith equation ($i = 2, 3, \ldots, n$). As a result we obtain an equivalent system where the first column of factors is equal to $\{1, 0, 0, \ldots, 0\}$. Then the second equation is divided by a_{22}, successively multiplied by a_{i2} and subtracted from the ith equation ($i = 1, 3, \ldots, n$). As a result, the second column of the factors for the derived system proves to be equal to $\{0, 1, 0, \ldots, 0\}$. Continuing this process, after the nth step we shall obtain an equivalent system with a unit matrix. At each step the diagonal elements are assumed to differ from zero. When system (6.9) is solved by the given method this assumption is fulfilled automatically.

It was found in experiments that after transforming the coordinates of points of an image corresponding to its rotation relative to the SV course, characteristics Δ, γ, S turned out to differ considerably from the reference values. This was established to result from a part of the image of a region on the planet staying beyond the limits of the readout system receptor field. Therefore to process large files and to provide for regular program operation in the transformation of coordinates corresponding to the image rotation, the following steps were taken. Instead of transforming formulae

$$
\left.\begin{aligned}
x' &= x \cos \alpha + y \sin \alpha \\
y' &= x \sin \alpha + y \cos \alpha
\end{aligned}\right\}
\tag{6.13}
$$

the following transformation was introduced:

$$
\left.\begin{aligned}
x' &= (x \cos \alpha) \times 10^{-2} + (y \sin \alpha) \times 10^{-2} \\
y' &= (x \sin \alpha) \times 10^{-2} + (y \cos \alpha) \times 10^{-2}
\end{aligned}\right\}
\tag{6.14}
$$

where α is the rotation angle of the image representing a region on the planet relative to the SV course. Moreover, the whole solution was shifted into the vicinity of a unity with the help of the second transformation of coordinates

$$
\left.\begin{aligned}
x'' &= (x' + 200) \times 10^{-3} \\
y'' &= (y' + 200) \times 10^{-3}
\end{aligned}\right\}
\tag{6.15}
$$

which is equivalent to certain permissible simultaneous transformations of the receptor field and the analysed image of the planet region.

The above method of data coding and remote transmission has shown its economic efficiency in data accumulation, transmission, and processing. It is important that in this method the Δ, γ, and S characteristics do not depend on SV flight direction relative to the regions, rotations, and linear displacements of images in the field of view of the SV readout system, which significantly simplifies the instrumentation and raises the reliability of obtained data.

6.1.2. Normalized maps for space navigation

The object of using normalized maps for space navigation is to raise spacecraft control reliability by means of the visual fixation of SV position relative to the terrain, i.e. by visual identification of images on the map with the objects on the planet surface, seen on board the SV, accompanied by simultaneous computer analysis of the normalized map image. The use of normalized maps is effective when it is necessary for man to take part in direct spacecraft control or in control monitoring.

A normalized map must meet the following requirements:

1. its appearance must closely approximate the appearance of the planet surface;
2. images modelling the planet surface should constitute the background of the map, and the specialized substantive elements should be represented in the form of strokes;
3. specialized elements of the map content, meant to be read out by a computer, can be visible, if they are to be simultaneously perceived by eye, or invisible, if this requirement is not imposed (i.e. when they are read and processed by computer only);
4. invisible data, intended for automatic reading, can be applied with luminescent paints—if they are read out by photoelectronic transducers, or with ferromagnetic paints—if they are read out by magnetic transducers (in principle, a map can be applied on ferromagnetic materials, making possible the recording and reading of data as on magnetic tape).

The reading head should be able to be positioned over any given section of the map, i.e. make possible the quasi-random access to information, necessitated by the choice of SV landing area or the fixation of SV position in the automatic control mode. A microscanner or a perception with a high density of retina receptors (matrix), i.e. with a sufficiently minute structure of the field of receptors ensuring the most comprehensive perception of the elements of cartographic image, can be used as a reading head. The reading head has to be easily manoeuvred manually over the body of the map.

Let us consider one of the possible types of a normalized map as exemplified by a planet surface typical for the Moon and Mars. For surfaces of this kind the basic substantive element is the relief. It is obviously expedient to represent the relief forms using the method of normalizing the isolines with a latent optical code (see sections 3.2 and 4.3).

Contour lines effectively represent the geometric characteristics of the relief surface, but they are inapplicable for out-of-scale objects: shallow craters, central uplifts, pits, etc., even if their height or depth exceeds the chosen contour interval. The prevalent forms of the relief of the celestial bodies in question are craters and, to a somewhat smaller extent, fractures, canyons, trenches, etc. It is expedient to represent such microforms by stroke symbols, with the use of the multiscale method (see Chapter 2), against a photographic (preferably coloured) background of the mapped celestial body surface. It is very important that the background image of the map should resemble the natural appearance of the celestial body surface. With multiscale representation the symbols of craters, fractures, etc. occupy a comparatively small part of the image space of the map. This leaves the background of the map sufficiently perceptible.

A great majority of craters are circular, or closely approximate a circle. Therefore in the multiscale method the craters unrepresentable by contours are conventionally represented not by a circumference but only by their diameter. The symbol of a crater is a line with its centre corresponding to the centre of the crater, and the length corresponding to its diameter. This symbol is taken as the basic one, to which are then added the characteristics of crater depth and profile. Differences in scale factors, characterizing the differences in crater diameters, can be shown by a different line thickness or colour hue (see section 2.5.3). To represent the shape (type) of a crater and its depth a special contour interval is taken, smaller than that for the main contours, proceeding from the average size of craters. Small craters are shown only by a line, i.e. without their profile and depth characteristics.

Fractures are best represented by the multiscale method, with the scale varying only along their width. The differences in scale factors can be shown by the differences in the luminosity of fracture image. Depths and other characteristics of fractures are best shown by strokes perpendicular to their direction (Fig. 6.1). Such symbols are comparatively easily and reliably identified by a computer, and are sufficiently easy to read visually.

On board a space vehicle the results of reading the portions of the map preassigned by the operator can be shown on the screen of the graphical display of a man–machine dialogue system in an enlarged scale to facilitate identification control and exert the necessary effect on control devices with the help of a light pen.

6.2. Digital Correction of Aerial Photographs for Relief Distortions

Stringent requirements on the accuracy of aerial photograph processing in automated systems necessitate elimination of distortions and reduction of photographs to a more invariant form of the orthogonal plan type. This is primarily dictated by the requirements of joint interpretation of photographs with one another and with cartographic material. Improving the equipment performance, as well as using analogue methods for photograph transformation, solve the problem only partially, since this does not adequately resolve the remaining question of eliminating the non-linear geometric distortions caused by the surface relief. These distortions manifest themselves in the radial directions, and at the photograph edges can be so great as to significantly narrow its effective areas.

The standard way of correction, using stereoscopic pairs or slot transformers, is not always acceptable. Digital methods of correction have recently found use for this purpose [26, 27].

Outlined below is one of the technically efficient methods of digital correction of geometric distortions on photographs, caused by surface relief, in which a digital model of the relief is used. Note that the digital relief model can be composed automatically, based on either normalized maps of isolines or traditional maps, after some preliminary preparation (see Chapter 4).

Let us consider the distortions caused by the local relief. We shall assume that all the other types of distortions have already been eliminated, and the photograph has been referred to the map of the terrain. Let us introduce a system of polar coordinates with the origin at the optical centre of the photograph. Photograph fixation permits writing in the same system the relief model function. Correction will be understood as the aerial photograph reduction to a form of the orthogonal plan type, since it does not depend on the point where the photograph was taken and the surface relief. Then, from geometric considerations, it is not difficult to obtain the correlation between the coordinates of points on the initial and the corrected photograph:

$$\bar{\rho} = \rho\left(1 - \frac{h(\rho, \varphi)}{H}\right)^{-1}$$

$$\bar{\varphi} = \varphi$$

(6.16)

where: $(\bar{\rho}, \bar{\varphi})$ are the coordinates on the initial photograph,
(ρ, φ) are the coordinates on the corrected photograph,
$h(\rho, \varphi)$ is the relief model function,
H is the altitude at which the photograph was taken.

Regarding the Earth as a reflector with a uniform indicatrix of diffusion, we can also write the equality of optical densities $\bar{P}(\bar{\rho}, \bar{\varphi})$ of the initial and $P(\bar{\rho}, \bar{\varphi})$ of the corrected photograph at respective points:

$$P(\rho, \varphi) = \bar{P}(\bar{\rho}, \bar{\varphi})$$

(6.17)

Generally speaking, mapping (6.16) may not everywhere be unambiguously reversible. We shall assume the surface relief and the altitude of photography to be such that the mapping of (6.16) is reversible. Reverse mapping can be constructed by means of the following iterative process:

$$\rho^{n+1} = \bar{\rho}\left(1 + \frac{1}{H} h(\rho^n, \varphi)\right), \quad \rho^1 = \bar{\rho}$$

(6.18)

Convergence of the iterative process is determined by the value of S:

$$S = \sup_{\rho} \left| \frac{\rho}{H} \frac{dh(\rho, \varphi)}{d\rho} \right| \qquad (6.19)$$

The condition of convergence, $S < 1$, is a sufficient condition for the reversibility of mapping (6.16). Usually the relief and the altitude of photography are such that three to five iterations are sufficient to achieve the required accuracy.

In the digital computer the image is represented in the form of a matrix whose values constitute the optical densities of the image at elementary discrete sites. The mapping of (6.16) makes it possible to obtain the corrected photograph by means of its 'global' point-by-point processing on a digital computer.

If the global point-by-point correction is unnecessary and all that is needed is to correct a certain structure on the initial photograph, one should resort to the reverse mapping of (6.18) and recalculate the coordinates of the points representing this structure.

When the above procedures are implemented on a digital computer, difficulties of processing a large quantity of data within the time available arise. Therefore, with the aim of increasing the rate of processing the exact transformation of a photograph is replaced by a similar, in a sense, transformation. The photograph is subdivided into rectangles, with the degree of relief variation taken into account. Only the corners of the rectangles are represented by exact transformation. From the data of the exact transformation of these corners a continuous transformation of the whole photograph is performed so that rectangles become angles. This transformation should be as simple as possible, without actually differing from the exact transformation.

Let us construct such a transformation. For this purpose we shall first consider a certain continuous and differentiable mapping of plane (x, y) into plane $(\bar{x}, \bar{y})$:

$$\left. \begin{array}{l} \bar{x} = FX(x, y) \\ \bar{y} = FY(x, y) \end{array} \right\} \qquad (6.20)$$

Let transformation (6.20) satisfy the following conditions:

1. The corners of the preassigned quadrangle in plane (x, y) turn into the fixed corners of the quadrangle of plane $(\bar{x}, \bar{y})$, i.e. the overall transformation is determined by the exact transformation of four points:

$$\left. \begin{array}{l} \bar{x}_i = FX(x_i, y_i) \\ \bar{y}_i = FY(x_i, y_i) \end{array} \right\} \qquad (6.21)$$

   where $i = 1, 2, 3, 4$.

2. The sides of quadrangles turn into sides, i.e. into segments of straight lines (note that exact transformation (6.16) does not, generally speaking, process this property):

$$\left. \begin{array}{l} FX(x_i + (x_{i+1} - x_i)t, \ y_i + (y_{i+1} - y_i)t) - \bar{x}_i = \bar{t}(\bar{x}_{i+1} - \bar{x}_i) \\ FY(x_i + (x_{i+1} - x_i)t, \ y_i + (y_{i+1} - y_i)t) - \bar{y}_i = \bar{t}(\bar{y}_{i+1} - \bar{y}_i) \end{array} \right\} \qquad (6.22)$$

   where $t, \bar{t} \in [0, 1]$, $i = 1, 2, 3, 4$.

3. The relation on the sides of the quadrangles is preserved:

$$t = \bar{t} \tag{6.23}$$

This condition is sufficient to eliminate image discontinuities on the sides of the quadrangles.

If a rectangle with sides hx and hy is subjected to transformation (6.20), then from (6.22) and (6.23) it follows that the transformation sought can be linear with respect to any one of the variables with the other one being fixed. Such bilinear transformation is determined by eight parameters and is sufficiently simple to calculate.

Considering (6.21), we obtain a transformation:

$$\left.\begin{aligned}
FX(x, y) &= \tilde{x}_1 + (\tilde{x}_2 - \tilde{x}_1)\frac{x - x_1}{hx} + (\tilde{x}_4 - \tilde{x}_1)\frac{y - y_1}{hy} \\
&\quad + (\tilde{x}_1 - \tilde{x}_2 + \tilde{x}_3 - \tilde{x}_4)\frac{(x - x_1)(y - y_1)}{hxhy} \\
FY(x, y) &= \tilde{y}_1 + (\tilde{y}_2 - \tilde{y}_1)\frac{x - x_1}{hx} + (\tilde{y}_4 - \tilde{y}_1)\frac{y - y_1}{hy} \\
&\quad + (\tilde{y}_1 - \tilde{y}_2 + \tilde{y}_3 - \tilde{y}_4)\frac{(x - x_1)(y - y_1)}{hxhy}
\end{aligned}\right\} \tag{6.24}$$

The errors of coordinates calculation, when (6.24) is used instead of (6.16), will be determined by quantities:

$$\left.\begin{aligned}
\Delta x &= \max_{x,y}\left[(hx)^2\frac{d^2}{dx^2} + (hy)^2\frac{d^2}{dy^2}\right]\frac{x}{1 - \dfrac{h(x, y)}{H}} \\
\Delta y &= \max_{x,y}(hx)^2\frac{d^2}{dx^2} + (hy)^2\frac{d^2}{dy^2}\frac{y}{1 - \dfrac{h(x, y)}{H}}
\end{aligned}\right\} \tag{6.25}$$

where Δx, Δy are the errors of coordinates calculation,
$\max\limits_{x,y}$ is the maximum for a quadrangle.

Expression (6.25) enables us to estimate the maximum dimensions of rectangles into which a photograph should be subdivided. The values of second derivatives are usually such that transformation errors do not exceed the errors of discretization for rectangles of appreciable size.

In this way, performing transformation (6.24) for every rectangle in the photograph, we transform the whole photograph. Condition (6.23) guarantees the continuity of transformation on the sides of rectangles, and the transformation of every rectangle does not depend in this case on the adjacent rectangles and is sufficiently easy to calculate. Moreover, a considerable part of the rectangles is transformed in accordance with simpler rules, special cases of (6.24). If the following equalities are valid:

$$\left.\begin{array}{l} \bar{x}_1 - \bar{x}_2 + \bar{x}_3 - \bar{x}_4 = 0 \\ \bar{y}_1 - \bar{y}_2 + \bar{y}_3 - \bar{y}_4 = 0 \end{array}\right\} \tag{6.26}$$

a rectangle is transformed into a parallelogram and transformation (6.25) can be actualized on a digital computer by a certain shift of lines and columns with omissions and additions. If, besides,

$$\left.\begin{array}{l} \bar{x}_4 - \bar{x}_1 = 0 \\ \bar{y}_2 - \bar{y}_1 = 0 \end{array}\right\} \tag{6.27}$$

a rectangle is then transformed into a rectangle, and computer implementation of transformation (6.26) becomes even simpler.

Note that the rectangles located in the vicinity of the photograph's optical centre are usually converted into themselves, and are not subject to transformation. Besides, certain quadrangles may happen to be filled with homogeneous brightnesses, and this also speeds up the transformation considerably.

The above algorithms have been realized on a digital computer. The input data for the program comprise a scanned aerial photograph; the data of referencing with respect to the terrain map; and the conditions in which the photograph was taken (altitude, camera orientation); as well as the digital relief model obtained from the normalized isoline maps. The scanned image is stored on a disk in the form of a random-access file. The program permits computation of images containing up to $2.10^3 \times 2.10^3$ points with 256 brightness gradations. According to the digital relief model, exact transformations of the corners of quadrangles are computed. Provision has been made for up to 200×200 points, which is quite sufficient for most of the photographs. Corrections are made for each of the rectangles by means of simplified bilinear transformations. The corrected image is also stored on a disk in the form of a random-access file.

6.3. Preparation of Original Cartographic and Aerospace Materials for their Automatic Conversion into a Digital Model

The most commonly applied technique involves converting the results of photograph interpretation or of instrumental surveys on a map manuscript into a digital form by tracing the image manually with the help of a reading device (coder) followed by the data input into a computer or recording them on a machine data-storage medium. This kind of cartographic data transformation requires large labour expenditures; at the same time the accuracy of conversion is not very high due to the errors associated with manual tracing. The necessity of automating this process is obvious. Labour productivity can be significantly raised if this complex problem is approached from the position of finding an optimal division of functions between man and machine.

Procedures associated with the identification of objects, i.e. data interpretation (particularly by image shapes and outlines) are in most cases performed more reliably by man than by machine; therefore it is expedient to assign a certain part of these functions to man. As to the interpretation results processing and map plotting, as well as identification by spectral characteristics, these can be realized by a machine in a fully automatic mode. In this case, however, all the procedures should be designed for automatic processing, beginning with depicting the image to be interpreted. This means that the photograph interpretation results should be graphically presented with due account for the resolving power and the logical potential of the available automatic devices and computer. For the input of interpretation results it is most expedient to use scanning devices or devices of matrix type (perceptrons) enabling simultaneous automatic input of the whole image. The technique in question presupposes the use of only scanning or matrix-type devices for the input of images into a computer, without the use of interactive devices, i.e. when the interpretation conditions make their use impossible, difficult, or inexpedient. This is so in the following cases: interpretation in the field; office studies during expeditions, when the use of interactive processing systems is impossible; mass processing of aerospace photographs, when the number of available interactive systems is insufficient for all the image decoders (operators) to be occupied; interpretation of complex objects consisting of minute elements and requiring a thorough tracing of the outlines of the object. It is also necessary to bear in mind that interactive processing systems for half-tone images are not cheap. In mass interpretation of aerial photographs it may prove much more advantageous to interpret the photographs directly, apply the normalized images on them, and then insert the data into a computer with the help of high-speed scanners. In this case subsequent processing, i.e. construction of digital models and map plotting, can be run in a fully automatic mode, with an interactive system used to control the course of processing. Interactive processing systems are especially advantageous in cases where the deciphered objects are difficult to distinguish visually by their colour tone or where for them to be distinguished it is necessary to synthesize several images obtained in different spectral regions.

The special features of presenting the interpretation results for subsequent automatic readout are as follows [124]:

1. In the course of interpretation the image can be directly applied on a photograph or on transparent plastic material placed over it. The latter is preferable, since it provides for a more reliable automatic readout of the interpretation results.
2. The boundaries of deciphered contours are drawn as an uninterrupted line of black colour. No restrictions are imposed on the quality of image, i.e. the image of deciphered objects can be applied by any specialist regardless of his draughtsmanship.
3. Qualitative characteristics of objects should be denoted by symbols in the simplest form (from the point of view of identification reliability and processing speed) for automatic readout. The symbols should differ in colour or structure. The choice of the colour for a symbol is determined by the type of scanning device to be used. If the readout is to be performed by

a single-channel device it is sufficient for the symbols to differ in the brightness of colour only. If a multichannel device (providing for a multi-coloured input into a computer) is used, the symbols can be of chromatic colours.

To increase the reliability of the automatic identification of symbols, when they are directly applied on a photograph, the method of latent optical coding is recommended. In this method luminescent paints with different emission spectra are used. At the moment of reading, the image drawn with these paints is irradiated by an ultraviolet source without being exposed to daylight illumination. This will result in the photographic image, be it coloured or black-and-white, becoming imperceptible to a computer, and only the luminescence light signals emitted by the image drawn on the photograph being recorded.

The reliability of automatic symbol identification will in this case also be much higher than for a conventional coloured image, because the emission spectra of luminescent paints are narrow. Furthermore, the reliability of automatic symbol identification can be increased by using luminophors with a different persistence constant. In a computer they can be distinguished not only by the wavelength but also by the duration of luminescence persistence.

When the deciphered image is applied on transparent plastic material, both conventional and luminescent paints can be used. Luminescent paints are, however, more effective in this case, providing for a better reliability of automatic symbol identification.

The use of multispectral paints is mainly expedient in drawing the symbols of linear objects. The form of unicoloured symbols recommended for linear objects, easily and reliably identified by a machine, is shown in Fig. 6.1. As to

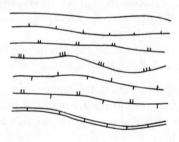

Figure 6.1

contour-like and discrete objects, it is recommended their qualitative characteristics be designated by digital code signs (DCSs), placed in the centre of a contour (Fig. 6.2a).

For DCS lines to be reliably identified by a machine, and easily applied manually, these lines should be taken sufficiently differing in thickness. If the signs are to be scanned by a device of universal type the difference in thickness between lines of different weight must be no less than two discretization

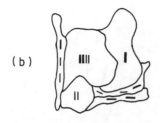

(b)

(a)

Figure 6.2

elements. To raise the reliability the difference in thickness can be increased. The thinnest line and the intervals between lines must also be equivalent to no less than two discretization elements. A DCS is not to contact or intersect the contour boundary.

When a processed map contains contours so small and narrow that the DCSs denoting them project beyond the contour boundary, another principle of DCS construction is used.

Let us consider the principle of DCS construction which offers a greater freedom in the modification of line combinations and simplifies the procedure of sign application. It should be noted, however, that this principle requires a more complex algorithm for the identification of such code signs.

In elaborating the principles of the DCS system described below the following requirements were taken into account: maximum information content, applicability to regions of any shape, simplicity of sign application, and identification by man. It is easily seen that these requirements are in many respects contradictory. The method considered below, as shown by practical investigations, constitutes a convenient compromise.

A DCS is given as a set of short lines (strokes) of arbitrary orientation and length. They can be freely arranged within the boundaries of a region (Fig. 6.2b). A line is identified by one parameter—its thickness. In practice only a few (three or four) classes of thickness are used, with certain errors in their assignment being permissible. To increase the reliability it is convenient to assume that the line length is more than twice its thickness.

When a DCS is identified, it is immediately assigned a certain integer, and further on all the subsequent data for the region are associated with it. The integer is obtained according to the following rule. Let us assume that scanning

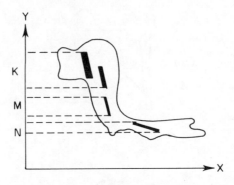

Figure 6.3

is performed parallel to the x axis. Projecting all the lines onto the y axis we shall obtain several segments—k, m, n (Fig. 6.3). We then enumerate the DCS lines in the following manner. Numeration starts with the strokes conforming to the upper segment, k, and proceeds from left to right. DCS is assigned the number containing at the kth place the thickness number of the kth stroke. For instance, in Fig. 6.3 the number assigned to DCS is 2111.·In this case the DCS is identified only by the outline of the contour. The contour boundaries are first traced by their extreme points (Fig. 6.4a), and then every DCS stroke is traced individually by its extreme points (Fig. 6.4b). To identify the weight of each stroke it is necessary to determine its length l and area S, and then the mean thickness, $d = S/l$. After this the weight of each stroke is found, proceeding from the preassigned criteria of thickness, and then the numerical value for the whole combination of strokes forming a code sign.

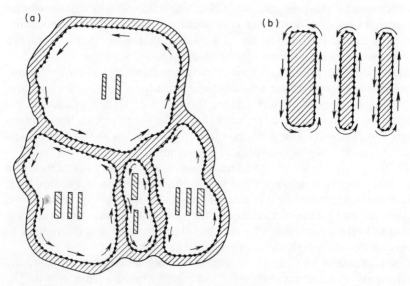

Figure 6.4

Let us now examine in greater detail one of the realized algorithms of DCS readout. In this algorithm all objects are processed uniformly; specifically, the areas are computed not only for DCS lines but also for all contour lines, since at the stage of tracing it is not clear which of the contours correspond to a DCS and which correspond to a boundary. The first thing to be done after tracing is to discard noises (image discontinuities, smudges), always present on a real map. Note that large and conspicuous noises can be eliminated manually at the normalization stage (inasmuch as, if they are present, they are small in number). As to small noises, especially when the quality of graphical information is poor, these may be present in considerable numbers. Naturally they should be disposed of automatically when processed by a computer. There are two kinds of noises: 'holds' (inner gaps) in a solid boundary and extraneous small dots. After the tracing both these types of noises are represented in computer memory by respective contours. If the area limited by a contour is smaller than a certain preassigned threshold value, this contour is regarded as noise and rejected. One should of course take care that this threshold should be less than the area of the minimum permissible DCS line.

Areas computed in the course of tracing have a sign—they are positive for contours traced clockwise and negative for contours traced counterclockwise. Note that DCS lines have two properties: firstly, their areas are positive; secondly, there are no other contours inside the contour of each DCS line. These two features make it possible to subdivide all the contours into two classes: contours of DCS lines and contours corresponding to boundaries. A special subroutine has been developed which allows determining for any two contours their relative positions (inside or outside). With its help one can define all the DCSs and determine the exact region they belong to.

Ascertainment of the class of thickness for every DCS line is based on the values of two parameters computed at the stage of tracing: half-perimeter P and area S. If the unknown values of line thickness and length are designated as d and l, from the obvious relationships of $l + d = P$ and $l \times d = S$ we can easily find d and, consequently, the thickness number. All the lines located within a region are ordered with the help of the found thicknesses.

6.4. Potentialities of the Integrated Interpretation of Remote Sounding Data and Maps in Solving Reclamation and Water Management Problems

The dynamism of water resources, closely interrelated with almost all the components of natural environment, their constant changeability under the impact of both natural and anthropogenic factors pose considerable difficulties in solving organizational, scientific, and technological problems of their management and control aimed at achieving their optimal distribution among water-users. A major problem is that of monitoring and recording the state of reclaimed lands over certain time intervals. This problem can today be successfully solved by using the methods of remote sounding, immediately followed by mapping. The advantage of these methods lies in the possibility of obtaining, within comparatively short time intervals and with a high economic efficiency, information concerning vast territories. This facilitates the solution of problems associated with the operation and control of territorial economic complexes and the monitoring of the state of reclaimed lands and water management systems.

Remote sounding methods can be particularly effective in resolving the questions connected with the rational distribution and utilization of water resources for the washing out of saline soils and the irrigation of arid agricultural areas where there is a shortage of water resources. It is well known that shortage of water substantially limits the possibilities of expanding irrigated agricultural land areas and increasing crop yields.

The problems of rational distribution of water resources cannot be solved in the absence of reliable and timely data on the condition of reclaimed lands, the actual state of storage reservoirs and the amount of water in them, on the state of main irrigation canals, drainage–collector network, etc. As shown by experience, all these data can be obtained with a certain degree of reliability and sufficiently high economic and technical efficiency by using the methods of remote sounding, provided that these data are interpreted jointly with the data obtained by other methods [15, 28, 30, 87, 91, etc.].

The most effective method of determining the quality of waters and their pollution is active sounding in the ultraviolet region with the help of laser radars (using nitrogen and helium–cadmium lasers). The method makes it possible to determine the area and thickness of an oil film on a water surface, the chemical composition of some pollutants (data taken from aeroplanes), the overgrowth of water reservoirs and streams, and many other phenomena.

With the help of radiometers in the microwave range (30–50 cm wavelengths), sideways-looking radiolocators, and radars with synthetic aperture (0.5–100 m wavelengths) one can determine depth to groundwaters, and in the 3–20 cm range one can determine the moisture content of soils and rocks in the aeration zone [28, 30, 87].

Multizonal surveys in the optical and near-infrared bands make it possible to estimate water quality in open watercourses and reservoirs (pollution with oils, phytoplankton concentration, etc.), to determine the state of their banks (destruction and overgrowth), conditions in the zones of influence of canals and storage reservoirs (rise of groundwater table, salinization, etc.), and many other phenomena.

Sounding in the red and infrared regions (within the spectral intervals of 0.7–0.9 μm in the daytime and 3.4–5.6 μm at night) appears promising for determining the humidity of the aeration zone surface layer. By measuring the water surface temperature, point sources of pollution are revealed, etc.

In multizonal surveys the open water surface manifests itself on aerospace photographs, taken in the red and near-infrared spectral regions, as a very dense optical tone. The ground surface with a heightened moisture content has a readily differentiable tone, more dense than the less moist soil. Because of this, water bodies and excessively moistened land areas can themselves serve as indicators in solving the problems associated with land reclamation. One of such problems is the determination of precise geometric parameters of relief for the levelling of reclaimed lands, and for feasibility studies in the preparation of reclamation and water management projects. For instance, when surveys are performed immediately after the spring snowmelt it is possible to determine photogrammetrically the exact geometry of the microrelief by the boundaries of inundated irregularities in land surface (hollows, ditches, pits, etc.). To identify the changes in the boundary of inundated irregularities it is expedient to perform several surveys in succession. Moreover, using as additional data the results of analysing the optical density of photographic tone, associated with the variations in the degree of soil moistening, one can increase the accuracy and detailedness of microrelief determinations.

When surveys are performed during the period of snowmelt it is possible to determine the state of indirect drainage, indicated by a different degree of snow melting, clearly represented on photographs and, in the absence of snow, by the distribution of soil moisture content.

Another serious problem where the optical characteristics of water surfaces and excessively moistened soils can serve as indicators, is that of revealing the sources of small streams and groundwater discharge for the purposes of their conservation and the estimation of water resources. In addition to this, by the spatial distribution and water discharge variations of the sources in different periods of the year, one can estimate the depth and thickness of groundwaters and their dynamics. Aerial photographs obtained in the long-wave optical and infrared regions make it possible to reveal the location of sources not only by the open water surface but also, in the case of subsurface sources, by excessive soil moistening and the presence of hydromorphic vegetation, represented on the photograph by an increased optical tone density. Phenomena, which are difficult for remote detection, and are encountered in reclamation, such as soil humidity and salinization, can be reliably interpreted only by combining different methods of sounding. Let us consider the example of determining soil moisture content, which is a very important characteristic for the management of water resources distribution, especially in the arid zones.

Difficulties of estimating soil moisture content are associated with the influence of vegetation cover, surface roughness, soil composition, humus content, salinization, and many other factors. Combining the methods and the electromagnetic spectrum regions should ensure maximum spectral differences between soils with different moisture content, both in the upper aeration layer and with depth, and eliminate completely, or diminish as much as possible, the effect of noise factors (exerting a negative effect on valid data). These methods and spectral regions include:

1. sounding (surveys) by the spectrophotogrammetric, television, multizonal, photographic methods in the red zone with wavelength $\lambda = 0.62$–0.72 μm and in the near-infrared zone with $\lambda = 0.7$–1.3 μm;

2. polarization method in the optical short-wave range with $\lambda = 0.40–0.58\,\mu m$ and, to a certain extent, in the red zone with $\lambda = 0.59–0.69\,\mu m$ at phase angles exceeding 70°;
3. infrared methods with IR imagers and radiometers in the region of $\lambda = 3.5–5.6\,\mu m$;
4. methods of microwave radiometry and, particularly, the method of active sounding with radars having a synthetic aperture in the microwave range with $\lambda = 3–20$ cm, and the deep penetration with the aim of detecting groundwaters with $\lambda = 30$ cm and more, up to decametric range.

Radiometric methods have the greatest capabilities in determining the moisture content and, perhaps, are the only methods from those listed above that make it possible to penetrate below the upper layer of the aeration zone. Due to the fact that thermal radio radiation is formed by the effective emitting soil layer, we can estimate the soil moisture content at a certain depth. Comparison of measurements performed in different spectral regions makes it possible to obtain the vertical moisture profile. The depth of wave penetration decreases with increased moisture content, and is inverse to absorption. Soil moisture measurements in the microwave range are negatively affected by such soil characteristics as temperature, surface roughness (microrelief), mechanical composition, etc., as well as by vegetation. The latter increases the emissive power and the radiation microwave brightness temperature in comparison with bare soils. To account for the effect of vegetation cover, corrections are introduced, based on the ratio between the height of vegetation and the wavelength in the range of $\lambda = 0.4–3$cm. The effect of cloudiness and water vapour in the atmosphere does not exceed 1–2K.

The joint use of several surveying methods in different spectral regions will make it possible to take into account numerous side-effects. Specifically, surveys in the optical long-wave (red) range using the polarization method, and in the near-infrared region, will allow one to account for the effects of the vegetation, humus content, salinization, mechanical composition of soils, etc. A very important factor is to carefully choose the time of the day when surveys are performed. Unlike the optical region, the infrared region makes surveying at night possible, and microwave surveying makes it possible to measure the moisture content with average resolution, practically irrespective of cloudiness. It should be remembered that thermosurveying, based on recording the ground self-radiation, in the daytime is performed under insolation, and at night in the emission mode, with the daytime gradients being greater than the night-time ones.

There exists a general regularity in the location of the indicative interval as a function of moisture content: with an increase in the remote receiver wavelength the indicative interval shifts into areas with higher moisture content values. This regularity is essentially non-linear. Nevertheless, a linear dependence does exist. For instance, the relationship between the emissive power in the microwave range (mainly determined by the surface dielectric constant and the zenith angle) and the soil moisture content in the interval of relative humidity values from 5 to 15 per cent, is of a linear nature. Different surveying methods can be best combined with the help of a flying vehicle (aeroplane, satellite, etc.) furnished with the necessary set of vehicle-borne equipment. For systematic moisture content monitoring on reclaimed lands, repeated surface soundings are necessary (to be performed, according to the estimates of reclamation engineers, once a week).

A major problem in land reclamation is forecasting the course of processes involving moisture content, salinization, groundwater movement, etc., after the reclamation and water management projects have been completed, or when these conditions are monitored in between successive soundings. Increased accuracy (reliability) of forecasts will permit conducting the remote surface soundings of conditions on reclaimed lands less often, and thus raise their economic efficiency. At the same time, solution of this problem will increase the reliability of forecasting the development and yield of crops. Of particular importance is to optimize water requirements and water consumption, using different methods for this purpose (among them the methods of determining the onset of 'stress' conditions for crops, arising with water shortage).

The problem of forecasting the humidity and salinization of soils can be solved with the help of deterministic and probabilistic models. As shown by experience accumulated in this branch of science, short-term forecasting is more effectively described by deterministic models and long-term forecasting by probabilistic–statistical models. Both these types of models require a large amount of diverse data.

For deterministic models it is necessary to have data on filtration parameters of soils, boundary conditions, and surface flow. A number of data needed to determine the filtration parameters (seepage intensity coefficient, dependence of the moisture permeability factor on moisture content, lower boundary of impervious layer, depth to groundwater table, etc.) can be taken from soil and soil-reclamation maps. Data on the initial moisture content in the aeration zone can be obtained by combining the methods of multizonal and polarization soundings and that of microwave sounding.

The data on boundary conditions include evaporation, rainfall intensity, transpiration, hydrostatic pressure, surface slope, hydraulic radius, Chezy's friction factor, and others. Let us consider them one by one [132].

1. To determine the evaporation rate it is necessary to have the following:
   (a) ground surface temperature, obtained by remote thermal sounding;
   (b) air temperature and humidity, taken from the weather data;
   (c) force and direction of wind, obtained from the weather data, the data of thermosounding, and the data on the relief, taken from a topographic map.
2. Rainfall intensity—from the weather data.
3. Transpiration—from the data of remote soundings in different optical regions and the thermal microwave region.

4. Hydrostatic pressure—from the relief on topographic maps, stereoscopic aerial photographs and precipitation data.
5. Surface slope—from the relief on topographic maps and stereoscopic aerial photographs.
6. Hydraulic radius—from the data of surveys in the near-infrared range and the data on topographic maps (concerning the areas and volumes of water bodies).
7. Chezy's friction factor—from the data of multizonal surveying and the data from topographic and forest maps (on extent of floodplain overgrowth, relief, etc.).

It is thus seen that most of the data can be obtained by remote sounding, processing cartographic materials and statistical hydrometeorological data. Field observations will only consist in relatively infrequent control measurements at individual points, mainly at the initial stage of forecasting.

Integrated processing of aerospace and cartographic materials is especially effective in solving spatial problems, for example, determining water surface areas, water mass volumes, and inundated area boundaries at high floods.

Let us examine a solution of this problem with the help of scanning devices and a computer. The principal operations for the automated processing of cartographic materials and aerospace photographs to obtain the above-mentioned parameters are as follows:

1. normalization of a duplicated copy or the negative of the relief of a topographic or a specialized map;
2. data input into a computer by a scanning device and identification of isoline values;
3. interpolation and construction of a digital model of the relief, H_{ij};
4. fixation of aerospace photographs on the map by the objects identified on the map and on the photograph;
5. input into the computer of aerospace photographs obtained in the near-infrared spectral region and correction of its image in accordance with the relief model and the identified objects;
6. selection of water surface points by their quantization levels;
7. tracing the line of the inundated area bank, z, i.e. determination of coordinates $x_i y_i$ for this area boundary; if the floodplain in some places is covered with forest or bushes to such an extent that the banks are not visible, these are found by the contours in the digital relief model and coordinates of the exposed part of the bank on the photograph;
8. determination of water surface area from formula (7.1) or (7.3);
9. determination of the mean watertable level, H, from the digital of relief matrix and the bank coordinates;
10. determination of the volume of water masses from the coordinates of flooded area boundaries, the mean watertable level, H, and the digital relief matrix, according to the formula:

$$V = \sum_{i=1}^{n} \sum_{j=1}^{m} (\bar{H} - H_{ij})l^2M^2; \quad (i, j) \in L \qquad (6.28)$$

where l is the distance in cm between the points of relief matrix on the map; M is the scale denominator of the map.

6.5. Some of the Methods for the Automatic Processing of Aerospace Photographs

6.5.1. Automatic tracing of water courses in hollows and ravines, of river channels, and other linear objects on aerospace photographs

The data concerning water courses on relief surface are necessary to solve numerous reclamation problems, particularly for the levelling of reclaimed lands, surface runoff investigations, and in many other cases. The data on river channels are necessary in estimating water storage and in geomorphological studies aimed at ascertaining the nature of their spatial distribution. Also determined in this case are such characteristics as the densities of their distribution with respect to length, area, sinuosity, direction of flow, etc. These characteristics can be represented as digital models or derived maps. These maps can then be processed and used to solve forecasting and predesigning problems. Determination of the above characteristics is associated with a large amount of measurements and computations, which necessitates the employment of automatic devices and computers. In order to obtain sufficiently reliable results of computer processing it is necessary to have comprehensive and authentic initial data. In investigating the water courses on relief surface the data obtained from aerospace photographs taken in the infrared region in spring or autumn are sufficiently representative. Such photographs can be taken by a multizonal camera from the channel recording in the near-infrared region and/or with the help of a multizonal scanner in the medium-infrared spectral region. As already noted, the optical tone density is markedly enhanced in the places where soils are overmoistened. These sites will evidently be in a certain way characterized by depressions in relief, primarily hollows, ravines, gullies, etc. The images of these are processed with the help of scanning devices and a computer by identifying the sought subsets in accordance with the quantization levels which correspond to the preassigned brightness levels of linear objects. Automatic processing includes the following operations:

1. input of the image into the computer;
2. finding the subset sought;
3. identification of the median line;
4. obtaining an ordered sequence of the coordinates of points [112].

The principle of image input by a general-purpose reading device is well known, so we shall not dwell on this operation. It will only be recalled that the data are stored in computer memory in a matrix form, i.e. as a regular network (grid) of points. The position of every point in the matrix is determined by rectangular coordinates in the form of line numbers $(j - 1, j, j + 1)$ and column numbers in the line $(\ldots, i - 1, i, i + 1, \ldots)$. Each point has its own code answering the levels of image quantization by brightness. Preliminary processing consists in identifying the subset related to the points of sought linear objects. We shall now explain the peculiarities of identifying the subset sought on the photographs. There are various ways of finding this subset. Note that the subset interpreted on an aerial or aerospace photograph has a different degree of image brightness (blackening), connected with insufficiently clear-cut outward characteristics of the objects on the terrain, their different reflectivities, the influence of relief, etc. All the factors complicate the identification of the sought subset. One of the simplest ways is to establish the values of quantization thresholds. Thus, at any image S, for its transformation f, and a certain numerical value (brightness code), we choose such a set of points $M_{s,p}$ where in all the points the brightness of the transformed image, $f(S)$, is not less than P. With two quantization thresholds, related by the condition of $P_1 \leq P_2$, a subset is established, complying with inequality $P_1 \leq f(S) \leq P_2$.

For the subset selection it is possible to use a matched filter in combination with quantization. If we assume that transformation f describes the image passing through the matched filter, which can be given in the form of a normalized cross-correlation function $\iint SR/(\iint S^2 \iint R^2)^{1/2}$ representing the image, S, and a reference image, R, this will allow us, by quantizing $f(S)$, to select the points of the image S corresponding to the reference image R_1. Note that what is selected here as a subset is not the image R but only a point or a small domain. We can take as a standard on the map a symbol from the legend and on the photograph—previously decoded similar objects.

Identification of the sought subset on the image is thus virtually reduced to assigning its function, equal to unity at the subset points and to zero at all the other points. As a result, in subsequent processing we shall have to deal with a set represented by a binary code. For the subset the code is a unity, for the background it is zero. The subset can be subjected to various operations, including the isolation of other subsets from it. Then the median line is determined, which is necessary for subsequent processing, and needed to determine the metric parameters of linear objects, etc.

The median line of an extensive object will be understood as a subset of the points of an object which have the following properties:

1. The presence of not more than two adjacent points at every point on the median line (with the exception of branch points of the object). This means that the subset of points in Fig. 6.5a is not a median line, but it can be obtained if we exclude from it, for example, the points marked by a crossed circle (see Fig. 6.5b). The adjacent points are shown in Fig. 6.5c by a circle with a dot.

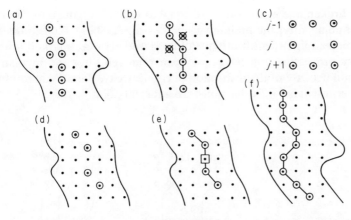

Figure 6.5

2. Connectivity. The subset in Fig. 6.5d is not a median line but it can be obtained by adding to the subset a point designated by a square in Fig. 6.5e. The distance between two adjacent points on the median line is thus equal to 1 or $\sqrt{2}$.

3. Replication by the median line of the outlines of the left and the right edges of a contour (see Fig. 6.5f).

The algorithm of median line identification consists of preliminary processing and the identification proper. At the jth step of preliminary processing a portion of the map comprising three lines, $j - 1$, $j + 1$, is treated; j is the processed line and $(j - 1)$ and $(j + 1)$ are auxiliary lines (the jth step implies the second scanning line, from which the analysis of the arbitrarily chosen portion of a photograph or a map begins). For every point of the jth line the sets of adjacent points are examined, and if in some of them there is even one background point, it is marked (e.g. its code is substituted by another one, established beforehand). All the lines are processed in this way.

After this we pass on to isolating the precise median line. At the jth step, as at the preliminary stage, a portion of the map consisting of three lines is treated. In the jth processed line all the points marked at the previous stage are examined in turn. If there is a non-zero point adjacent to the point examined, or its removal disrupts the connectivity of non-zero neighbouring points, the point remains, i.e. its code is substituted by the one it had prior to the preliminary processing. Otherwise the point is removed, i.e. its code is substituted by the background code or some other preassigned code.

Let us specify what is meant by the disruption of the connectivity of non-zero adjacent points (the points that at a given moment belong to the object). Removal of a point disrupts the connectivity of adjacent points if it results in a situation when it becomes impossible to join all the adjacent points with a

214

connected broken line whose sides are equal to 1 or $\sqrt{2}$. Figure 6.6a shows the position of points after the preliminary processing, and Fig. 6.6b shows them after the median line identification. Points II and III were marked after the preliminary processing and have not been removed after the median line identification because of the disruption of connectivity. Point IV has not been removed because it has only one non-zero adjacent point.

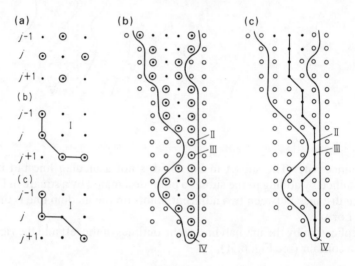

Figure 6.6

In the course of median line identification the points removed at this stage are counted. If their number is not equal to zero, preliminary processing is reverted to and the whole procedure repeated. Equality to zero means that the median line has already been traced, i.e. either every point has one adjacent point or its removal disrupts the connectivity of adjacent points. Note that the position of the median line does not depend on the scanning direction, as the data are stored in computer memory in the form of a regular network of points, and the position of every point in the network is determined by its rectangular coordinates. The proposed principle of median line identification is equally suitable for all continuous linear objects, irrespective of the shape of curves and their spatial position relative to the scanning lines.

The points having only one non-zero adjacent point are terminal (located at the ends of linear objects). Their coordinates are identified separately and serve as the starting points of tracing. In Fig. 6.6b such a point is denoted by IV. For a closed contour, any point on the median line can be the starting point; it is also the final point where the tracing ends.

The procedure of obtaining an ordered sequence of points for linear objects is based on a simple algorithm: point a_i, nearest to the last tracked point a_{i-1}, is determined. The closest distance is

$$\left.\begin{array}{l} |a_i - a_{i-1}| = d \\ \text{or} \quad |a_i - a_{i-1}| = d\sqrt{2} \\ \text{if } d = \Delta x = \Delta y \end{array}\right\} \qquad (6.29)$$

where: Δx is the line advance of the scanner,

Δy is the discretization step along a line.

In most reading devices Δx and Δy are equal to each other. Every subsequent point is thus considered as the nearest one only if the distance to it from the last point coincides with one of the above parameters. As a result of tracing, an ordered sequence of coordinates for the points on the axial lines of objects is established. The described method of determining the median (axial) lines is applicable to any continuously represented extensive object regardless of its width, curvature, and length.

6.5.2. Automatic determination of the degree of floodplain overgrowth with forest vegetation

Determination of the degree of floodplain overgrowth with forest vegetation is necessary in surface runoff calculations when studying the hydrological regime of rivers and designing reclamation and water management structures. This problem is specifically associated with determining the necessary parameters of forest vegetation to find the roughness factor which enters as an argument into the well-known formula of Chezy used in average flow-rate calculations.

The roughness factor is directly related to the degree of floodplain overgrowth. The principal parameters of the latter, as shown by numerous investigators, can be determined with sufficient reliability and effectiveness by means of remote sounding methods, in particular aerospace photography, with the results (photographs) interpreted visually or with the help of automated optical and electronic equipment. To solve this problem automatically, using scanning devices at computer input, it is necessary to formalize the method of image interpretation on aerial photographs. A large number of works have been devoted to the problems of interpreting (deciphering) forest vegetation for a great variety of purposes (with the exception, perhaps, of the problem in question) using visual and automatic methods. Analysis of the experience gained shows that the problem in question can be successfully solved by automated methods; consequently it yields itself to formalization and computer processing.

The basic parameters of forest vegetation to be taken into account when determining the degree of floodplain overgrowth, are the average distance between trees, and average height and species composition of forest stand within a uniform section of a forest. In determining the average height of trees, and distance between them, the distinctive features are the structure and tone (optical density) of an image. In determining the species composition it is the combination of tone parameters on multizonal photographs or colour parameters on spectrozonal photographs. The most obvious way to solve this problem on a computer is by using the methods of pattern (image) recognition that have been realized in practical forest mapping with a certain degree of processing efficiency. Let us consider this technique, proceeding from the presently known recognition methods. Its realization involves the following successive general operations:

1. digital representation of a photographic image in computer storage;
2. identification of features and their classification;
3. finding the decision function;
4. representation of recognition results in cartographic form.

The first operation should be performed with the help of photoelectronic devices providing multigradational quantization by the image brightness levels. The result of the input of aerial photograph data into a computer is a digital regular matrix of readout points imposed on the image field at intervals usually equal to the discretization step. Such a form of representation is standard and therefore its further consideration is unnecessary.

The second operation presupposes devising indicative algorithms necessary to form from the elements of digital image certain quantitative characteristics closely associated with the parameters of interpreted objects. In the case of composing maps that characterize the average distance between trees, indicative features are represented by the gradations in forest stand density, and the distinctive features are represented by the structure of the image formed by tree crowns. This structure, transformed in computer storage (as a result of scanning and input), will be characterized by the alternation of over- and undershoots of digital quantized matrix values relative to a preassigned threshold. Undershoots correspond to image areas lighter than the tone of the threshold, overshoots correspond to darker areas. The recurrence frequency of these shoots by itself characterizes the forest stand density and should be regarded as one of the classes of features. Uniformity of areas is determined by the statistics of over- and undershoots. To identify the features representing spatial and frequency characteristics of image area, associated with forest stand density, it is necessary to have an algorithm for determining the recurrence frequency of over- and undershoots [137]. The recurrence frequency of undershoots at the ith threshold can be determined from the following formula:

$$\left.\begin{aligned} f_i &= \frac{1}{L} \sum_{m=0}^{M-1} \sum_{n=0}^{N-1} \theta_{mn}^{(i)} \\ \theta_{mn}^{(i)} &\begin{cases} 1, & \text{if } D_{mn} - D_i < 0 \text{ and } D_{m,n-1} - D_i \geq 0 \\ 0, & \text{in all other cases} \end{cases} \end{aligned}\right\} \quad (6.30)$$

where L is the length of the scanning line expressed by the number of elementary sections; M is the number of lines; N is the number of elementary sections in a line; D_{mn} is the current value of optical density for an elementary section σ with integer coordinates of the centre m, n; D_i is the value of optical density at the ith level.

In processing multizonal photographs it is necessary to have an algorithm to identify a feature by the multidimensional average tone. In different zones one and the same area of the terrain will be determined by a combination of its average tones. A multidimensional tone carries information on the colour of the image, and therefore on the reflective properties of the surface of the object shown on the photograph.

Reduced integral illumination intensity of the light-sensitive layer in the receiver of the ith zone is expressed by a well-known relationship:

$$E_i(x, y) = \int_0^\infty b(\lambda, x, y)\tau(\lambda)\tau_{0i}(\lambda)\varphi_i(\lambda) \, d\lambda \qquad (6.31)$$

where $b(\lambda, x, y)$ is the spectral luminance of the object, $\tau(\lambda)$ is the spectral transmissivity of the atmosphere, $\tau_{0i}(\lambda)$ is the spectral transmission coefficient of the optical system and the filter for a given zone, $\varphi_i(\lambda)$ is the spectral sensitivity of the receiver.

Reduced luminance of an image being dependent on the spectral characteristic of the filter in a particular zone, the average tone of one and the same object will be different in different zones. That is why every object can be represented by the k-dimensional vector of tone in each zone.

The third operation is associated with automatic attribution of the image area with the value of identified feature to certain class. Methods of pattern recognition can be applied for this purpose. The procedure of solving this problem is reduced to constructing reference blocks in computer storage that include training sequences of image parameters of the typical forest areas for which all the necessary data are available. By processing an image of an area we obtain the vector of its parameters. If a forest area is represented by a set of parameters forming an n-dimensional vector, then every class S, of a feature in parametric space will adhere to n-dimensional partition law $W_k(x_i, \ldots, x_n)$ of the coordinates of vector x. The input data for the decision algorithm will be represented by the identified vector $X = (x_i, \ldots, x_n)$. The decision algorithm, depending on the relationship between the elements of the vector and the reference vectors, makes one of m decisions, $\gamma_1 \ldots, \gamma_m$, as to which one of m classes $S_1, \ldots, S_m$ the identified vector belongs.

An important element of this problem is the finding of optimization criteria for decision algorithms. Different approaches and techniques can be used for this purpose.

To determine the similarity between objects α and β a measure of closeness, d, is used as a function of distance in parametric space. This function is found from formula

$$d_{\alpha,\beta} = \sum_{i=1}^n k_i(x_{i\alpha} - x_{i\beta})^2 \qquad (6.32)$$

where k_i is the weight or normalizing coefficient.

The fourth operation, performed with the help of cartographic algorithms, pertains in principle to the already solved problems briefly dealt with in Chapter 5.

CHAPTER 7
Automation of Measurements and Investigations on Maps

7.1. Automation of Measurements on Maps

The problem of obtaining certain quantitative characteristics of objects from maps (lengths of lines, areas, volumes, etc.) pertains to the application of maps. Application of maps is a broad concept. It includes data acquisition from maps (with cartometry as a part of it) and the analysis of the data obtained.

Before modern means of automation found their use in cartography, data acquisition from maps had been performed visually with the help of simplest measuring devices such as a rule, dividers, measuring grid, or curvimeter. With the appearance of computers and reading devices the approach to data acquisition from maps has radically changed, and cartometry as a part of cartography has actually lost its previous independent importance. What is essential today is the automatic reading of maps, including the readout of cartographic information, its computer input, and the identification of specified objects with the aim of determining their quantitative parameters. The object of such reading of maps is to analyse data in performing all kinds of investigations and solving different practical problems. These problems are most effectively solved in a fully automatic mode with the use of scanning readout devices and mini- or general-purpose computers.

Automation of the acquisition of measured cartographic data is thus closely associated with computerized reading of maps. Acquisition of measured data is not the ultimate aim in most cases. The results of measurements are later processed. For instance, the results of measuring the lengths of rivers are used in plotting the maps of river network density to be used in ecological, hydrological, geomorphological, and other investigations. Determination of areas is necessary for correlation, factor, and other methods of analysis, and determination of volumes is necessary to estimate water and mineral resources, to determine excavation volumes when designing canals, roads, etc. This processing is now performed in a fully automatic mode. Because of the integrative nature of the operations performed and their laboriousness it is expedient to perform cartographic works with the help of mini- or general-purpose computers. It is obviously unnecessary to construct an autonomous specialized device with its own processor for such uses of cartographic information. There may, however, be some specific exceptions.

The second reason why it is expedient to perform cartographic measurements with the help of a reading device interacting with a general-purpose computer is the fact that automatic acquisition of measured data from maps using a specialized device is a very complicated matter. As is known, the most complicated factor is the recognition of symbols. To perform the recognition functions without a general-purpose computer it is necessary to develop a rather complicated specialized computing device having, as it may turn out, very modest functions restricted within the framework of solving individual problems only. Because of the narrow scope of the problems solved such devices can become economically inexpedient. In field conditions the use of conventional and automatic planimeters, curvimeters, etc., where image recognition and contour-line tracing are performed manually by man, is still justified. It is well known how intricate, painstaking, and long is the procedure of outlining a contour manually. What is also very important, manual measurements cannot assure a high accuracy. That is why, for mass measurements in laboratory conditions, it is expedient to automate all these processes by means of general-purpose computers and scanning devices.

The methods for determining the metric parameters, examined in this chapter, are meant to be implemented with the help of scanning devices and computers. In principle they can also be used to determine the metric parameters with simple devices such as measuring grids. In this case the scanning lines formed by the reading device can be regarded as the lines of a measuring grid and the discretization points along the lines as the nodes of this grid.

7.1.1. The measurement of areas and perimeters of contours

Numerous methods and devices exist for the measurement of contour areas and lengths of lines at a differing level of automation. Complete automation of the measurement of areas and lengths is attained with the help of scanning devices and computers [87, 107, 110].

Two methods are recommended for the automatic measurement of areas and perimeters of contours. The image of contours is read out by a scanning device. The reading device has the sole function of putting information into a computer. All the other operations are performed by the computer.

The first method is based on the principle of calculation from coordinates, and the second is based on the principle of a dotted grid.

The process of measuring the areas and the perimeters of contours by the first method
This consists of the following operations:

1. readout of contour images and their input into a computer;
2. tracing the contours, i.e. determining an ordered sequence of coordinates of points for every contour;
3. calculating the areas and perimeters of contours from coordinates;
4. output of the results of computer processing in graphical and digital form for the total and individual area and perimeter values.

Let us now consider the principle of readout and computer input for contours represented on a map as continuous boundary lines, which is perhaps the most typical form of representing contours on a map.

Taking into account the fact that these contours have no differences in brightness, they can be regarded as two-gradation images. As no analysis of the image by the levels of its brightness is required the image is immediately transformed into a two-gradational one. The points belonging to the image are assigned the (01) code and those of the background the (00) code. The position of points in plan is specified by the ordinal of the line and the ordinal along the line.

Let us turn to the second operation—the tracing of contours, directly connected with computer processing.

The tracing starts with searching for the first initial point. As soon as its coordinates are fixed (let them be (i, j) in Fig. 7.1a) the point with coordinates $(i - 1, j)$ is taken for the initial, and with (i, j) for the central one, and all the points are then checked in the sequence 1–2–3–4–5–6–7–8, shown in the figure. When a point with a code differing from $(0, 0)$ is encountered, this point $((i - 1, j + 1)$ in Fig. 7.1b) is taken for the central one, and the following point (i, j) for the initial one, and all the points are again traced according to the scheme shown in Fig. 7.1. Every newly found point (its coordinates or ordinals, which is the same) is stored in computer memory. Every point is compared with the first initial one (i, j) (Fig. 7.1a) and, if they coincide, a marker of the end of the tracing (two successive zeros) is inserted. To exclude repeated tracing of the contour every point is compared with $(1, 0)$ and, when they coincide, code $(1, 0)$ is replaced by $(0, 1)$. After the tracing of the next contour the map is examined along the lines from left to right, and the tracing stops when the lower right-hand point is reached. It should be borne in mind that if contours have

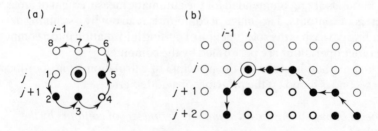

Figure 7.1

common boundaries the tracing is performed along the internal points of contour boundary.

Other methods of tracing the contours can also be used (see, e.g., section 4.4).

Let us now consider the third operation: calculating the areas and perimeters of contours.

The area of a contour can be easily determined from its coordinates:

$$S = \sum_{i=0}^{n-1} (y_{i+1} - y_i) \frac{x_{i+1} + x_i}{2} \tag{7.1}$$

where $y_{i+1}, y_i, x_{i+1}, x_i$ are the orthogonal coordinates of contour points; n is the number of points along the internal or external contour line.

The perimeter of the contour is determined from interpolated points lying midway between the nodal points, using formula:

$$L = \sum_{i=1}^{n=1} \sqrt{(y_i - y_{i+1})^2 + (x_i - x_{i+1})^2} \tag{7.2}$$

Total values of contour areas and perimeters are found by simple input of the ordinals of the contours to be summed up.

Finally, the fourth operation of the output of processing results onto a printer is in the form of the coordinates of extreme points (to designate the measured contour) and the corresponding areas. The results can also be represented as schemes of contour boundaries with the indication of the area inside every contour.

The process of area measurements in the second method
This consists of the following principle operations:

1. readout of contour images and their input into a computer;
2. selection (identification) of points on the contours with the recoding of zero points falling inside the contour into a code assigned to this contour;
3. calculation of the area of every contour;
4. output of results onto a printing device.

The first operation is similar to that of the first method. The second operation consists in line-by-line processing of points. As a rule two lines are processed, starting from the line passing through the extreme point. By comparing the ordinal numbers of neighbouring points we establish their belonging to a certain contour. The algorithm of this operation has been presented in detail in [107] and [110].

After selecting all the points filling the contour their area is calculated from formula:

$$S_n = N_k d^2 \tag{7.3}$$

where N_k is the number of points filling contour k; d is the scanner discretization step, i.e. the distance between discretization points ($d = \Delta x = \Delta y$). This method, as seen from the formula, is based on the grid principle. Unlike the first method it does not permit determining the contour perimeter.

The accuracy of measurement is governed by the size of the scanner discretization step. Technical errors of the scanners are negligible. The principle of accuracy estimation is the same as for the grid. Using well-known approaches to estimating the grid accuracy and the dependencies suggested to determine mean square errors, one can easily estimate the accuracy of the methods in question. Taking into account the fact that the discretization step for existing scanners can amount to $0.2 \div 0.05$ mm (which means that the scale factor of the grid will be equal to $0.04 \div 0.0025$ mm^2, respectively), we can expect in advance that the accuracy will be higher than the graphical accuracy of the map.

A general-purpose computer can be attached to the scanner as the computing block.

To provide for a complete automation of area measurements on prepared maps it is necessary that the contours to be measured should be selected (recognized) by their colour or by the form of symbol images. This procedure of computer recognition can prove to be difficult for maps and charts carrying a large amount of graphical information. In this case it is possible to determine all the contours on the map without selection and, after the output of computed results, to choose the necessary ones. The contours drawn as continuous lines of black colour are reliably determined. The presence of other chromatic and grey-colour images on the map does not in this case affect the reliability of identification and the accuracy of measurements.

The performed measurements of areas have shown the high technical and economic efficiency of this method.

Measurements with the use of these methods are only possible in laboratory conditions, requiring more expensive equipment than the conventional planimeter and curvimeter; therefore the latter instruments will not lose their importance in the future in the practice of measurements.

Analogue and digital circuits of electronic planimeters are used in the calculation of contour areas.

Analogue circuits are based on the use of phase-sensitive scanning circuits whose advancing pulses are fed into the inlet of a quadratic detector with an amplitude modulator actualizing the dependence

$$S_c = 2 \frac{v}{v_0} m \times q \qquad (7.4)$$

where m is the modulation depth (constant); q is the ratio between the displacement of the beam on the screen of the tube and the increment of the integrator potential causing this displacement; S_c is the area limited by contour c; v_0 is the initial potential of the scanning integrators; v is the final potential at the outlet of the quadratic detector.

Digital circuits are based on the use of discrete line scanning circuits and of discrete signals coming from the scanning circuit along the x- and y-axes to actualize, with the help of two reversible counters and an accumulator, the dependence

$$S_c = \sum_i y_i \Delta x_i \qquad (7.5)$$

where y_i is the value of the ordinate at the ith step of scanning; Δx is the increment of the abscissa at the ith step of scanning.

7.1.2. Methods of determining the areas of non-planar surfaces, represented on a map with isolines (contour lines)

The area of a non-planar surface has to be determined in solving various problems. It is of particular interest in determining the irrigated areas for perennial vegetation and agricultural crops planted on slopes, etc. The area of a surface represented by isolines, for example of relief on a topographic map, can be determined with the help of a scanning device and a computer in different ways. The choice of the method depends on the type of the original map, i.e. whether it is fully normalized or traditional.

Let us first consider the methods for a fully normalized map produced as described in reference 111 and section 4.3.1, making it possible to determine isolines unidimensionally, i.e. from one scanning line. It is expedient in this case to use the method based on the summation of profile lengths multiplied by the scanning step [114, 120].

$$S = \tfrac{1}{2} \sum_{j=1}^{m} (L_j + L_{j+1})d \qquad (7.6)$$

where L_j is the length of the profile along the scanning line within the region boundaries; d is the step of scanning, $d = \Delta x = x_{j+1} - x_j = (b - a)(m - 1)$; m is the number of scanning lines within the measured area, i.e. from a to b; j is the ordinal number of the scanning line ($j = 1, 2, 3 \ldots m$).

The profile length is found by interpolation between the intersection points of the scanning line with the isolines. In the simplest case it can be calculated from the formula:

$$L_j = \sum_{i=1}^{N} \sqrt{l_{j,i}^2 - \Delta Z_{j,i}^2} \qquad (7.7)$$

where $l_{j,i}$ is the distance between the two neighbouring intersection points of the scanning lines and the isolines, $L_{j,i} = Y_{j,i} - Y_{j,i-1}$; n is the number of intervals along the line; i is the ordinal number of the interval ($i = 1, 2, 3, \ldots, n$); $\Delta Z_{j,i}$ is the contour interval value, i.e. the difference between the values of neighbouring isolines intersected by the scanning line, $\Delta Z_{j,i} = Z_{j,i} - Z_{j,i-1}$.

Formula (7.6) yields satisfactory results under certain conditions. For instance, when finding the area of a topographical surface the scanning lines should go perpendicular to the orographic lines, such as the directions of valleys, ravines, etc.

224

Formula (7.6) does not take into account the changes in area associated with the sloping of the surface whose slope vectors are perpendicular to the scanning lines. To eliminate these errors the scanning should be repeated in the direction perpendicular to the initial one and then the following formula should be applied:

$$S = \tfrac{1}{2} \sum_{j=1}^{m} (L_j + L_{j+1})d + \sum_{k=1}^{p} (L_k - L_k^{(0)})d \tag{7.8}$$

where L_k is the profile length within the region boundaries obtained by the second scanning; $L_k^{(0)}$ is the projection length of the profile obtained by the second scanning within the region boundaries; p is the number of the second scanning lines; k is the ordinal number of the line of the second scanning ($k = 1, 2, \ldots, p$).

When constructing digital models of relief from traditional maps (see Chapter 4), or from the previously obtained digital model stored in the data bank, it is expedient to determine areas from the points of a matrix. In this way a greater accuracy of area determination can be achieved. Generally speaking it is necessary to calculate in this case a double integral

$$\iint_S f(x, y) \, dS = \iint_S \sqrt{1 + \left(\frac{dz}{dx}\right)^2 + \left(\frac{dz}{dy}\right)^2} \, dy \, dx \tag{7.9}$$

of a function of two variables, $z = f(x, y)$, taken over the area S of the given surface. Its numerical solution consists in the summation of elementary areas obtained from a regular rectangular grid (matrix) of points.

It should be noted that the denser the points of the matrix the smaller the size of elementary areas, and therefore the higher the accuracy of measurements. Numerical solution of this problem can be found with the help of vector analysis. Figure 7.2 shows an elementary area restricted by four neighbouring

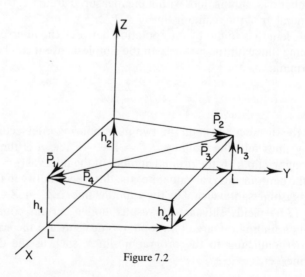

Figure 7.2

points of a digital matrix (grid). In accordance with the designations taken in Fig. 7.2 it can be written that

$$
\left.
\begin{aligned}
\mathbf{P}_1 &= L\,\mathbf{l}_x + \mathbf{l}_z\,(h_1 - h_2) \\
\mathbf{P}_2 &= L\,\mathbf{l}_y + \mathbf{l}_z\,(h_3 - h_2) \\
\mathbf{P}_3 &= L\,\mathbf{l}_x + \mathbf{l}_z\,(h_3 - h_4) \\
\mathbf{P}_4 &= L\,\mathbf{l}_y + \mathbf{l}_z\,(h_1 - h_4)
\end{aligned}
\right\}
\tag{7.10}
$$

Hence the elementary area of a surface can be expressed as

$$
\begin{aligned}
S &= \tfrac{1}{2}\left|[\mathbf{P}_1, \mathbf{P}_2]\right| + \tfrac{1}{2}\left|[\mathbf{P}_3, \mathbf{P}_4]\right| = \\
&= \tfrac{1}{2}L^0\sqrt{L^2 + (h_1 - h_2)^2 + (h_2 - h_3)^2} + \\
&\quad + \tfrac{1}{2}L^0\sqrt{L^2 + (h_3 - h_4)^2 + (h_1 - h_3)^2}
\end{aligned}
\tag{7.11}
$$

Proceeding from the condition of a plane $([\mathbf{P}_1, \mathbf{P}_2], \mathbf{P}_3) = 0$ we can write that

$$
h_1 + h_3 = h_2 + h_4.
\tag{7.12}
$$

As a result, the formulae for determining the elementary area of a surface, restricted by four points of a regular grid with the values of h_1, h_2, h_3, h_4, and the total area of the surface will acquire the following form:

$$
\left.
\begin{aligned}
s &= L\sqrt{L^2 + (h_1 - h_2)^2 + (h_2 - h_3)^2} \\
S_0 &= \sum_{j=1}^{m-1} \sum_{i=1}^{n-1} L\sqrt{L^2 + (h_{ij} - h_{i+1,j})^2 + (h_{i+1,j} - h_{i+1,j+1})^2}
\end{aligned}
\right\}
\tag{7.13}
$$

where L is the distance between the points of the matrix; i is the ordinal number of a point along the line $(i = 1, 2, \ldots, n)$; j is the ordinal number of a line $(j = 1, 2, 3, \ldots, m)$.

7.1.3. Corrections for scale differences in map projection in area measurements

It is expedient to measure areas from maps in equivalent projections. In these projections the areas are known to remain undistorted. When the areas are measured on medium- and small-scale maps plotted in other projections it is necessary to introduce corrections.

The problem of correcting the measurement results for scale differences of maps has been considered by numerous authors. But these questions have been considered as applied to the simplest measurement means. The methods of measurement with simple appliances and with computer-assisted scanning devices have certain distinct differences. Let us illustrate the peculiarities of accounting for distortions caused by scale differences as exemplified by the two most widely used projections: Mercator conformal projection and Gauss–Kruger conformal transverse cylindrical projection [120].

The scale of area P_s, equal to the square of the linear scale in Mercator projection, is known to be determined from formula

$$P_s = m^2 = \sec^2 \varphi \qquad (7.14)$$

where φ is the latitude of the site.

In this projection the general scale is only preserved along the equator. The scale being a function of latitude only, it remains the same along any of the parallels. This feature of the projection indicates the expediency of the scanning to be performed along the parallels. It is obvious that on small-scale maps with parallels in the form of arcs it is impossible to scan along the parallels as the scanning lines are as a rule straight. This non-parallelism, however, will not seriously affect the accuracy of area calculations. Because of the compensation of errors we can in practice assume that the scale of an area within a narrow band between two lines limiting the area, ΔS, remains constant.

In order to take into account the scale differences in area calculations it is necessary to fix beforehand the latitude, φ, of the upper point of the boundary (from which the scanning starts) for the area to be measured and the increment, $\Delta \varphi$, corresponding to the scanning step. After this the corrected value, $\Delta S_j^{(0)}$, can be calculated from formula

$$\Delta S_j^{(0)} = \frac{\Delta S_j}{\sec^2 (\varphi - j\Delta\varphi)} \qquad (7.15)$$

where j is the ordinal number of the scanning line.

Formula (7.15) is applicable to a trapezoid lying above the equator. For trapezoids lying below the equator the denominator should contain the following expression:

$$\sec^2 (\varphi - j\Delta\varphi)$$

When calculating areas from maps represented in Gauss–Kruger conformal transverse cylindrical projection the direction of map scanning should be parallel to the x axis, since in this projection the constancy of scale is preserved along lines parallel to the axial meridian. The corrected value of the increment of area, $\Delta S_j^{(0)}$, is found similarly to the case of Mercator projection:

$$\Delta S_j^{(0)} = \frac{\Delta S_j}{\left[1 + \frac{1}{2}\left(\frac{y + \Delta yxj}{R}\right)^2\right]^2} \qquad (7.16)$$

where Δy is the scanning step; R is the Earth's radius; y is the coordinate of the upper point of determined area, D_0.

7.1.4. Methods of calculating the volumes from isolines and digital models

Calculation of volumes is necessary to solve various problems of scientific and industrial nature. Volumes have to be determined in numerous cases: to calculate the amount of excavation necessary for the construction of canals, roads, the levelling of reclaimed lands, etc.; to estimate water masses in storage reservoirs, lakes, etc.; to calculate atmospheric precipitation or thickness of glaciers in determining the available water supply; to ascertain the necessary water diversion zones in preventing the harmful effect of water upon agricultural lands, hydraulic and water management structures, etc.; to calculate the available mineral resources, and to solve many other problems.

Depending on the specific problem to be solved different maps plotted in isolines or aerial photographs can be used, but in actual practice most often used for such purposes are topographic maps containing detailed information on the surface relief.

Measurements and calculations can be performed both with the help of simple appliances and on a computer [114]. The methods presented here involve the use of scanning devices and computers. When choosing the necessary technique one has to take into consideration first of all the type of the original (the processed map), the required machine time and accuracy. The original can be in the form of a normalized map or a traditional map requiring partial normalization (see Chapter 4). The differing degree of noise in the representation of isolines (i.e. the presence of images not related to isolines) should be taken into account.

Let us consider three methods of volume calculation:

1. from horizontal sectional planes;
2. from vertical sectional planes;
3. from points of a regular digital matrix.

Any surface represented by isolines constitutes a continuous function, $f(x, y)$, in an enclosed area, D, restricted by the internal frame of the map or by an external boundary of arbitrary form. The volume can be presented as the following expression:

$$V = \iint_D f(x, y) \, dx \, dy$$

Isolines by themselves belong to the continuous–discrete method of representation where horizontal section $\Delta Z = $ const. (with a few exceptions) and section $Z = Z_i$ is of arbitrary form. In accordance with this method, horizontal sectional planes are used to calculate the volume by summing up parts, of the volume ΔV, restricted along coordinate Z_i by horizontal sectional planes with a constant step, ΔZ, represented on the map by traces of neighbouring isolines. The partial volume can be represented by formula

$$\Delta V_i = \frac{S_i + S_{i+1}}{2} \Delta Z \tag{7.17}$$

where S_i is the section area; the total approximated volume will then be equal to

$$V = \sum_{i=1}^{n} \Delta V_i = \tfrac{1}{2} \sum_{i=1}^{n} (S_i + S_{i+1}) \Delta Z \tag{7.18}$$

where i is the ordinal number of the section $(i = 1, 2, 3, \ldots, n)$.

In practice, however, most often—especially when working with a relief map—the volume is determined within boundries of specified area D and section Z_0. The prescribed value of the section, Z_0', as the starting point of readout, may not coincide with any isoline value. It can, for example, be the design elevation of constructed canals, or buildings, specified in accordance with optimal project data. It can only accidentally coincide with an isoline. Furthermore, to obtain a more accurate value of the volume it is necessary to take into account the surface extrema unrepresentable on a map (summits, pits, etc.). These are usually expressed as the volume of a cone whose base corresponds to the area restricted by the isoline at the extrema and preserved as an increment to the general formula. These increments can be both positive and negative.

With all this taken into account the formula of the volume will acquire the following form:

$$V = S_0'(Z_1 - Z_0') + \Delta Z \left[\frac{1}{2} \sum_{i=1}^{n} (S_i + S_{i+1}) + \frac{1}{3} \sum_{k=1}^{m} S_k \right] \qquad (7.19)$$

where Z_1 is the value of the first isoline starting from Z_0' along a specified direction; S_0' is the area of the assigned region; k is the ordinal number of an extremal isoline ($k = 1, 2, 3, \ldots, m$); m is the number of extremal isolines; S_k is the area of the section along the extremal isoline. A series of complex computer operations is required to determine the area of the section, S_i, which can be limited by a trace of one isoline or by several isolines having the same value. First the computer must identify (select) all the intersection points of the scanning lines with the isolines by their values and then trace them, i.e. arrange their coordinates in an ordered sequence. After that the areas of these sections are calculated from the ordered coordinates. If the area of the section is not completely limited by the trace of an isoline (because it projects beyond the region boundaries), it is determined from boundaries of area D. The area is determined from formula (7.1).

Let us now examine the principle of determining the volumes with the help of vertical sectional planes.

In the case when area D is limited at the base by a rectangle, i.e. at the top and the bottom it is limited by straight lines $y = c$ and $y = d$ and on the left and the right by $x = a$, $x = b$, with the continuous function $f(x, y) \geq 0$ throughout area D, the volume will be equal to

$$V' = \int_a^b dx \int_c^d f(x, y) \, dy$$

From this it follows that the main problem of measurement is reduced to determining area $S(x)$ of the vertical section by the plane $x = \text{constant}$,

$$S(x) = \int_c^d f(x, y) \, dy \quad \text{(Fig. 7.3)}.$$

Note that in this method, as distinct from the preceding one, determination of a vertical section area constituting a curvilinear trapezoid is much simpler.

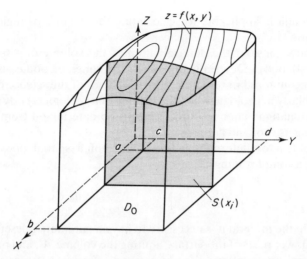

Figure 7.3

Since isolines are identified in a unidimensional space (along one scanning line) it becomes possible to calculate volumes on a computer concurrently with data input. We shall now examine this in greater detail.

Computer reading of a normalized map of isolines makes it possible to calculate the volume concurrently with the scanning along individual lines. Calculation of volumes by individual lines is meant as determining increment ΔV_j when the isolines on a map are crossed by a single scanning line. Having summed up all the ΔV_j we obtain the sought volume.

Increment ΔV_j, resulting from the passage of one scanning line across the isolines on a map, can be represented by the following formula:

$$\Delta V_j = \Delta x S(x_j) \tag{7.20}$$

where $\Delta x = x_{j+1} - x_j = [(b - a)/m] = d$; m is the number of scanning lines within the range from a to b ($j = 1, 2, 3, \ldots, m$);

$$S(x_j) = \Delta y_{0,j} Z_{0,j} + \Delta y_{1,j} \tfrac{1}{2}(Z_{1,j} + Z_{2,j}) + \cdots + \\ + \Delta y_{n-1,j} \tfrac{1}{2}(Z_{n-1,j} + Z_{n,j}) + \Delta y_{n,j} Z_{n+1,j} \tag{7.21}$$

where n is the number of intersection points of the scanning beam (scanning element) with isolines within the boundaries covered by one line; $Z_{0,j}$ and $Z_{n+1,j}$ are the values at the intersection points of the scanning beam with the boundary of area D, assumed to be equal to $Z_{1,j}$ and $Z_{n,j}$, respectively.

Proceeding from the aforesaid we can write the formula for the whole volume:

$$V = \sum_{j=0}^{m} \sum_{i=0}^{n} \Delta x [\Delta y_{ij} \tfrac{1}{2}(Z_{i+1,j} + Z_{ij})] \tag{7.22}$$

where i is the ordinal number of an intersection point of the scanning line with the isoline ($i = 0, 1, 2, \ldots, n$).

This formula is applicable for calculating the volume of region D of any configuration.

Let us now consider the determination of the volume of a spatial (three-dimensional) body. Determination of the volumes of bodies is a frequent problem encountered in research in various fields of the science of the Earth, and particularly in geology, when the volume of a geological body (e.g. a water mass in estimating water resources) has to be determined from the isolines representing its roof and floor.

The problem of determining the volume of a spatial body consists in calculating a complex integral

$$\int_V dV = \int_a^b \int_{\varphi_1(x)}^{\varphi_2(x)} \int_{\psi_1(x,y)}^{\psi_2(x,y)} dz \, dy \, dx$$

To calculate the integral it is necessary to have a parametric description of the upper and lower parts of the surface limiting the volume, V, and presented in a triple integral by equations $Z = \psi_1(x, y)$ and $Z = \psi_2(x, y)$, respectively. In geology such surfaces are usually represented with a map of the isolines of the floor and a map of the isolines of the roof of the geological body. When such maps are available, volumes can be determined with the help of scanning devices and a computer by two methods.

For both methods, however, it is necessary to have a common reference point from which the readout of Z starts; i.e. a common hypothetical datum plane. Formulated in this way the problem of volume determination will consist in finding the difference between the hypothetical volume calculated from equation $Z_2 = \psi_2(x, y)$ for the roof surface and the volume obtained from equation $Z_1 = \psi_1(x, y)$ for the floor surface.

The most expeditious way to determine the volume of a body lies in line-by-line data processing on two maps simultaneously. In a scanning device the map of the isolines of the roof is successively fixed (in direction of scanning), and then the same is done with the map of the isolines of the floor, with the origin of x-coordinates of both maps exactly coinciding and the y-coordinates being displaced arbitrarily but in parallel.

With the scanning line passing along both maps the increment of the volume of the body is immediately determined, and then the sum of the two increments is found according to the formula:

$$V = \sum_{j=1}^m \Delta x [S^{(k)}(x_j) - S^{(n)}(x_j)] \tag{7.23}$$

where $S^{(k)}(x_j)$ and $S^{(n)}(x_j)$ are the areas of vertical sections of hypothetical bodies along the scanning lines for the roof and the floor, respectively.

Simultaneous processing of two maps in succession along the lines is possible with the help of a special scanning device. When using general-purpose scanning devices this method of measurement may prove difficult. In this case the hypothetical volume, $V^{(k)}$, is calculated from the map of the isolines of the roof according to formula (7.22), and then in the same way from a similar formula the hypothetical volume, $V^{(n)}$, is found on the map of the isolines of the floor. The sought final volume is calculated as the difference of hypothetical volumes

$$V = V^{(k)} - V^{(n)} \tag{7.24}$$

Determination of the volume with the help of a regular digital matrix having a constant distance between the nodes is performed from the same formula as in the case of calculations with a measuring grid:

$$V = \Delta S \sum_{i=1}^{n} (Z_i - Z_0') \quad \text{or} \quad V = S_0' \times \overline{Z} \tag{7.25}$$

where $\Delta S = \Delta x \times \Delta y$; Δx, Δy are the distances between nodes of the matrix along the x- and the y-axis, respectively; Z is the value of a point on the matrix; i is the ordinal number of a point on the matrix; n is the number of points on the matrix, falling within region D; S_0' is the area of region D; $\overline{Z}$ is the average value of points on the matrix within the area of region D,

$$\overline{Z} = \frac{1}{n} \sum_{i=1}^{n} (Z_i - Z_0')$$

A regular digital matrix can be obtained as a result of processing both a normalized and a traditional map, using one of the methods of model construction described in Chapter 4.

Determination of volumes from normalized maps with the help of scanners and computers is simpler and more expeditious when performed in accordance with the method based on vertical sectional planes, and from traditional maps it is simpler using horizontal planes and a digital matrix.

7.2. Methods of Constructing Maps and Digital Models in Investigations of the Spatial Distribution of Objects

Spatial distribution of extended objects, such as rivers, tectonic faults, hollows, or ravines, can be abstractly described by means of integral transformation of their images on maps and photographs in accordance with their basic geometric parameters and represented as digital models and maps characterizing in a generalized way their density with respect to extension, area, sinuosity (for rivers), degree of branching (mainly for ravines), as well as their statistical density, directionality, extensiveness, etc.

With the help of such formalized descriptions it is possible to estimate spatial distribution of objects and—having a retrospective description—to estimate their development. Furthermore, they allow one to conduct various investigations automatically as the form of their representation is easily perceived by a computer (as digital models on computer data media or in the form of normalized maps). Extensive objects on aerospace photographs can be transformed automatically if they are reliably identified by their spectral characteristics. Thus, for example, the water surface of rivers is represented on a photograph in the long-wave region of the optical spectrum and especially in the near- and medium-infrared region by an optically very dense phototone, which makes it possible to identify it easily by the levels of video signal quantization. From photographs taken in these and other regions of the electromagnetic spectrum one can also identify ravines, hollows, etc.

7.2.1. Method of constructing digital models and maps describing in a generalized way the density of linear objects by their extension and area (as exemplified by a river network)

Let us first consider the method of constructing digital models and maps representing the extensiveness of objects. The gist of the method consists in describing the distribution of the density of image lines by their total length and its representation in the form of isolines or raster lines of different frequency and thickness in the form of digital models, etc. The density of linear objects on a certain area is understood as the ratio of the total length of all the lines shown within a region to the area of this region [108].

The process of automatic map plotting consists of the following basic operations:

1. The readout of linear data from a photograph or a map by a scanning photoelectronic device and their computer input.
2. The processing of input data with the aim of obtaining a uniform network of points representing the values of line density, with respect to their total length, in the form of a matrix.
3. The recording of a digital matrix or a map at computer output as a linear raster or isolines.

The first operation is performed by a reading device for conventional computer input.

The second operation, associated with computer data processing, includes solving the following problems:

1. Determining the optimum dimensions, shape, and mutual spatial position of elementary areas.
2. Determining the approximate generalization coefficient, when the source material is of medium or small scale.
3. Calculating the density values for each elementary area and attributing them to their centres.

The general solution of the problem consists in dividing the source material into equidimensional elementary areas and finding within them the total density value as the ratio of the sum of line lengths to the area. After this the isolines are plotted.

The choice of the optimum size of the elementary area is determined by the problem conditions. The larger the area, $S_{i,j}$, the more generalized will be the boundaries of regions characterizing a certain density. It is also clear that overlapping elementary areas will convey the information more reliably than adjacent or separated ones. The greater the overlapping the more accurate is the description, but also the longer is the computer time. In general the size of the elementary area should be calculated individually for each concrete case, mainly taking into account the degree of dispersivity of the objects on the source map or photograph. The more rarefied the objects the larger the elementary area to be taken.

It is expedient to take the shape of the area as a square, as this speeds up and simplifies computer processing. For the same reasons the size of elementary areas and the degree of their mutual overlapping should be assigned as equal for all the areas. The degree of the overlapping should not exceed 50 per cent, as a further increase in the percentage of areas overlapping does not result in any noticeable increase in the accuracy of density description, but prolongs computer processing time. The principle of choosing the size and shape of the elementary area is the same for all the methods described in the section.

The coefficient of generalization for medium- and small-scale source materials is determined with the help of large-scale maps or photographs, where linear objects are shown with a minimum of generalization. This problem can be solved using a photograph or a large-scale map as an analogue. A region is selected on the source material, corresponding to the boundaries of the taken elementary area; it is then read out, inserted into a computer, and processed. The processing consists of line-by-line scanning of the selected region stored in computer memory in two directions: along the x- (Fig. 7.4a) and the y-axis (Fig. 7.4b). The number of 'encounters', n'_x and n'_y, of the scanning lines with the images of rivers is counted in accordance with the assigned optical density quantization level on the photograph. In Fig. 7.4 the points of encounter are shown by solid circles.

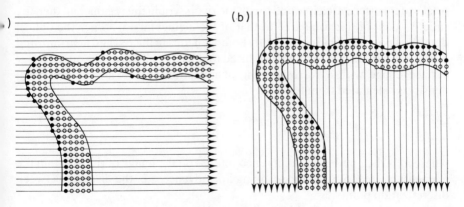

Figure 7.4

A similar procedure is performed on the source material of a smaller scale; the values of n_x, n_y are found for an identical region, and the generalization coefficient is then calculated from the formula:

$$g = \frac{n'_x + n'_y}{m(n_x + n_y)} \tag{7.26}$$

where m is the scale factor.

The values of elements P_{ij} of matrix $\mathbf{P}$ for each elementary area S_{ij} of the source material are calculated from the following formula:

$$P_{ij} = \frac{\pi g d}{4S_{ij}} (n_x + n_y) \tag{7.27}$$

where d is the scanning line advance; S_{ij} is the size of the elementary area; n_x, n_y are the numbers of points where scanning lines encounter the image within the limits of the elementary area.

If the discretization step along the x-axis differs from that for the y-axis in the reading device, the calculation is performed from the following formula:

$$P_{ij} = \frac{\pi g}{4S_{ij}} (d_x n_x + d_y n_y) \tag{7.28}$$

where d_x and d_y are the discretization steps along the x- and the y-axis, respectively. The obtained P_{ij} values relate to the centres of all the elementary areas. This results in a uniform network of points of matrix $\mathbf{P}$ with numerical values of density P_{ij} as its elements.

Finally, the last operation of the output of transformation results from the computer is well known.

Digital models and maps representing the density of objects dispersed over an area are constructed following the same principle. The original is divided into equidimensional elementary areas. From every such elementary area the total water surface area is found for all the rivers within the region, and then the ratio is taken of this total area to that of the elementary S_{ij}. Since the optical density of photographs is sufficiently homogeneous in the red and near-infrared bands, the problem of readout is reduced to selecting all the points on the water surface image with a corresponding quantization level. The value of areal density is calculated from formula:

$$P_{ij}^{(s)} = \frac{n_{ij} d^2}{S_{ij}} \tag{7.29}$$

where n_{ij} is the number of points encountered on the image of rivers within the elementary area, S_{ij}; d is the scanning line advance, i.e. the distance between discretization points.

This technique is applicable to all dispersed areal objects irrespective of their shape.

7.2.2. Construction of digital models and maps describing in a generalized way the distribution of objects according to their degree of branching, direction, and elongation (as exemplified by a ravine network)

One of the important characteristics of a network of ravines, making it possible to estimate the degree of erosion (along with the description of their density distribution in area and length) is the degree of branching, or dissection of ravines [120]. To plot the isolines and construct digital models characterizing these features of a ravine network all the mapped territory is divided, as in the previous section, into equal elementary areas S_{ij} having the shape of a square, and then within every square the q_{ij} values of the **Q** matrix, characterizing the degree of branching, are found from the following formula:

$$q_{ij} = \frac{1}{n_{ij}} \sum_{k=1}^{n_{ij}} \frac{L_k^2}{S_k} \tag{7.30}$$

where L_k is the perimeter of ravine k whose weight centre lies within the boundaries of S_{ij}; S_k is the area of ravine k within S_{ij}; n_{ij} is the number of ravines within S_{ij} ($k = 1, 2, \ldots, n_{ij}$). All the obtained values relate to centres of elementary areas. These values are then used to interpolate and draw the isolines.

Perimeters and areas are determined on a computer in accordance with the procedure described in section 7.1.

The following technique is recommended to describe the degree of elongation and the direction of a ravine network.

All the connected points of the specified subset (points covering the ravine) are regarded as a random field of points governed by a two-dimensional normal distribution law

$$f(x, y) = \frac{1}{2\pi\sigma_1\sigma_2} \exp\left\{-\frac{1}{2}\left[\frac{(x-a)^2}{\sigma_1^2} + \frac{(y-b)^2}{\sigma_1^2}\right]\right\} \tag{7.31}$$

The parameters of the dispersion ellipse ($a, b, \theta, \varepsilon$) approximating the ravine are found, where a, b are the ellipse semi-axes; θ is the slope of the major axis of the ellipse with respect to the x-axis; ε is the eccentricity.

The degree of ravine elongation can be expressed by formula

$$\varepsilon' = \frac{a}{b} - 1 \tag{7.32}$$

The directionality of a ravine is determined by angle θ. Based on these data a rose-diagram is plotted for every S_{ij}. This is done by finding the

$$\sum_{i=1}^{n_{ij}} \varepsilon'$$

values lying within discretely specified values of the rose-diagram angles in accordance with the direction of θ_k, and then plotting the obtained totals along these directions.

7.2.3. Construction of digital models and maps describing in a generalized way the degree of the sinuosity of extended objects (as exemplified by a river network)

The method described is that of plotting a map of the sinuosity of linear objects with the help of scanning devices and a computer [120].

A meander (element of sinuosity) will be regarded as a curve limited by inflection points (or, in other words, the points of zero curvature). The number of meanders will be taken equal to the number of inflection points.

The problem to be solved is that of finding within the boundaries of elementary area S_{ij} the values of sinuosity, a_{ij}, which are the elements of **A** matrix, and then using these elements to plot a map of isolines or a raster map. As in the preceding paragraphs, it is necessary to divide the source into a network of equidimensional rectangles (areas). The sinuosity values within the elementary areas are found from formula

$$a_{ij} = \frac{N_{ij}}{\Sigma \, L_{ij}} \tag{7.33}$$

where N_{ij} is the number of inflection points within S_{ij}; $\Sigma \, L_{ij}$ is the total value of the lengths of all rivers encountered within S_{ij}. The $\Sigma \, L_{ij}$ value is found from formula (7.2) after tracing the rivers, preferably by channel points (see section 6.5.1). The lengths should be calculated and the inflection points found not from the coordinates of grid nodes, obtained as a result of tracing, but from the average values between the nodes. From the obtained array of points the inflection points are found in accordance with the technique described in section 5.5.

7.2.4. Construction of maps describing in a generalized way the areal density of dispersed objects

Let us consider the principle of reading the cartographic information concerning dispersed objects and of plotting density maps with the help of differentiation and integration methods, as exemplified by a river network [120]. The principal problem is reduced to determining the Z_{ij} elements of **Z** matrix for the areal density of a river network. Figure 7.5 shows a scheme of data readout and conversion when the areal density map is plotted by means of a differentiation method. The data on source map 1a is represented as described above (see section 2.5.5). Map 1a is scanned by line scanning block 1 and read out by photoelectronic head 2 having a small scanning element $\Delta p = 0.02 = 0.04 \, mm$ (Fig. 7.6). At the outlet of block 2 (Fig. 7.5) signals of different voltage amplitude, $u_1, u_2, \ldots, u_i$, are formed (Fig. 7.6b). These are transformed by sync pulse generator 3 and analogue-to-digital converter 4 into a series of equidistant pulses with the amplitudes corresponding to the accepted quantization level (Fig. 7.6c). As an illustration, Fig. 7.5 shows a planar view of a map portion 3a after the conversion. It is obvious that the quantization levels and their amount should be chosen in optimal accordance with the image brightness gradations on the map. The data converted at the outlet of block 4 into digital

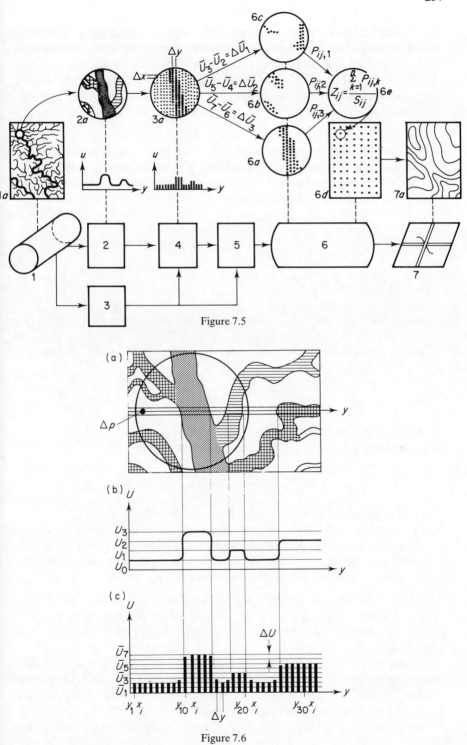

Figure 7.5

Figure 7.6

form (binary code) enters block 5 and then computer 6. Planar coordinates are stored in computer memory as the ordinal numbers of points (including those of the background) and the values of quantized brightness levels for every point (see Fig. 7.6c). The procedure of map readout and computer input ends here. Then the computer processing of the recorded data starts, aimed at plotting a derived map of areal density. The processing consists of the following procedures:

1. Selecting the points by quantization levels $u_1, u_2, \ldots, u_n$ in accordance with the accepted gradations on the map (Fig. 7.5, 6a, 6b, 6c; Fig. 7.6c).
2. Counting the number of points, N_{ij}, and then determining the area, for each gradation separately, from formula

$$\left. \begin{array}{l} P_{ij,1} = \dfrac{1}{K_1}\, N_{ij,1}\Delta x \Delta y; \quad P_{ij,2} = \dfrac{1}{K_2}\, N_{ij,2}\Delta x \Delta y; \ldots; \\[2ex] P_{ij,n} = \dfrac{1}{K_n}\, N_{ij,n}\Delta x \Delta y \end{array} \right\} \qquad (7.34)$$

where Δx is the distance between the points along the x-axis, equal to the scanning line advance, d; Δy is the distance between the points along the y-axis, equal to the discretization step prescribed by the pulse generator.

3. Calculating the total area,

$$\sum_{k=1}^{n} P_{ij,k}$$

occupied by linear objects within the limits of S_{ij}.

4. Determining the values of areal density elements,

$$Z_{ij} = \sum_{k=1}^{n} P_{ij,k}$$

and relating these values to the centres of elementary areas S_{ij}.

5. Linear interpolation by the Z_{ij} values at the centres of elementary areas and recording the results on plotter 7 in the form of a map of isolines, 7a. The centres of elementary areas are taken regularly (6d). It is advisable that elementary areas should overlap one another.

The results can also be recorded on an alphanumerical printer. In this case zones of equal values are covered with appropriate signs chosen from a previously plotted scale. The scale of Z_{ij} gradations can be represented in percentages.

Let us now consider the method of readout and the plotting of derived areal density maps with the help of the integration method. Figure 7.7 shows the scheme of data readout and transformation when an areal density map is plotted by the integration method.

Source map 1a is scanned by scanning block 1 with photoelectronic reading

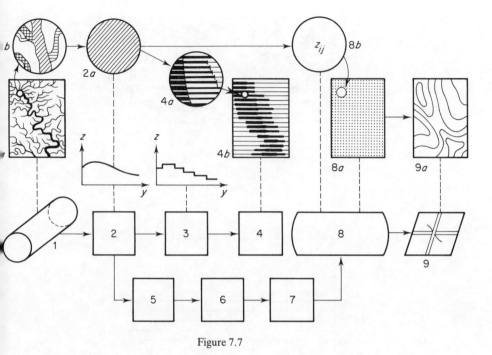

Figure 7.7

head 2 having a large scanning element. The size of the element can vary within rather wide limits, associated with the field of vision of the receiver lens.

The lens provides for the variation in the area of the field of vision from a few mm^2 to a few cm^2 on the map. The size of the field is preassigned, depending on the density of objects on the map, and its purpose. Figure 7.7 shows a portion of map 1b covered by the lens field of vision. The lens operates in a defocused mode, serving only to transfer the light energy from the map onto the photocathode of light-sensitive element 2a. In the course of scanning a continuous signal of varying voltage arises at the outlet of block 2 (Fig. 7.8b). Variation in the amplitude of the $Z_{x_i}(y)$ curve characterizes the density of linear objects within the elementary area when it 'slides' along the scanning line (Fig. 7.8a). Block 2 can have two channels at its outlet. Through one of them the data enter block 3 (multilevel pulse-height discriminator) quantizing the continuous signal by its amplitude. Quantization is performed in accordance with a specified scale of Z_{ij} gradations (Fig. 7.8c). The quantized $Z_{x_i}(y)$ signal enters a specialized recording block 4. Block 4 can record images in the form of a raster, 4a, 4b, or dotted isolines.

The second channel of block 2 (Fig. 7.7) is connected with computer 8 through a block of signal interruption 5, analogue-to-digital converter 6, and interface block 7. The values of Z_{ij} are input into the computer in digital form (8a, 8b). The computer performs only one operation: interpolation for the plotting of isolines 9a.

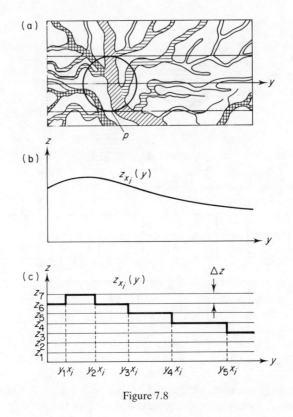

Figure 7.8

7.3. Determination of Interrelations of Phenomena from a Set of Maps

A set of maps representing different characteristic features of one or several phenomena (related to the same investigated region) enables their comprehensive analysis. Such analysis is primarily aimed at ascertaining the interrelations between, and the laws governing, the distribution of phenomena and objects in space and time. On traditional maps such investigations are performed visually or with the help of the simplest measuring devices.

Computers are now used to perform a large quantity of calculations. Their use, however, does not make things much easier; the processing of cartographic data still takes a lot of manpower and time. This is primarily associated with the fact that on traditional maps the information is represented in a form not convenient for computer readout and identification, which restricts the use of computers as the data are taken from maps and input into a computer manually. The difficulties of analysis are also caused by many types of information being mainly of a qualitative nature, combined with quantitative characteristics. This kind of information is most often represented on a map by enclosed connected areas, i.e. as contours. The quantitative method of data representation by isolines is also widely applied. Such a variety of the forms of data representation causes considerable difficulties in solving the problem of their simultaneous processing when conducting scientific investigations, especially when the problem of automatic processing arises. In this connection it has become necessary to develop new methods of analysis.

The most complicated is the development of techniques aimed at determining the interrelation between phenomena represented on a map by means of different characteristics and different representation forms and, what is more, having a different nature of spatial presentation, e.g. on one map continuous and on another continuous–discrete. An especially difficult case occurs when on one map the data are presented as a two-dimensional surface, i.e. as enclosed connected areas (contours) and on another in the form of a three-dimensional surface, i.e. by isolines. An idea of solving this problem, i.e. one of its possible solutions, has been suggested in references 116 and 120.

In those cases when the phenomena represented on a map are characterized only qualitatively, ascertainment of the closeness of the interrelation between them is only possible by the mutual spatial position of the boundaries of their contours and areas. The simplest analysis is based on evaluating the areas of the intersecting regions of the correlated phenomena. If a coordinated network of points is superimposed on one map and then the other, it becomes possible to estimate the degree of closeness of interrelation between phenomena by the number of points falling within the correlated regions of these maps and by the number of coinciding points, resulting from the intersection of these regions. Every point in this case can be regarded as an elementary area whose sides are equal to the distance between the network nodes. It is obvious that the more dense the network of points the more reliable is the determined value for the degree of closeness of interrelation between phenomena.

The indicator of the closeness of interrelation can be calculated from formula:

$$\rho_{a,b} = \frac{1}{2} \sum_{k=1}^{m} S_k(c) \left(\frac{1}{\sum_{i=1}^{m'} S_i(a)} + \frac{1}{\sum_{j=1}^{m''} S_j(b)} \right) \qquad (7.35)$$

where $S_i(a)$ is the sum total of elementary areas lying within the ith contour of correlated phenomenon a; $S_k(c)$ is the sum total of elementary areas, lying within contour k formed by the intersection of phenomena, $c = a \cap b$; i, j, k are the ordinal numbers of the contours of correlated phenomena a, b, and the areas of their intersection, respectively.

The problem becomes more complicated when one phenomenon is assigned on one map in the form of contours and on another as isolines. The map of isolines is then subdivided into zones. These zones can correspond to the contour interval taken for the map or chosen arbitrarily. Every zone will be regarded as a connected area in a two-dimensional space limited by isolines. Every such area will have an external and an internal boundary. The area can be multiply interconnected and therefore have several internal boundaries. All the points of internal and external boundaries are, however, conventionally regarded as the points of one and the same level of the Z value, and the interval, ΔZ, between the neighbouring areas is assumed to be constant. Having transformed the three-dimensional surface into a series of hypothetically planar two-dimensional surfaces we find the indicator of the closeness of interrelation, ρ, for each zone from formula

$$\rho_i(Z) = \sum_{k=1}^{m} S_k(c) \Big/ \sum_{i=1}^{m} S_i(Z) \qquad (7.36)$$

where $S_i(Z)$ is the sum total of elementary areas lying within the ith zone of phenomenon Z, represented on the map by isolines; i is the ordinal number of the zone ($i = 1, 2, \ldots, m'$); $S_k(c)$ is the sum total of elementary areas of contour k, formed by the intersection of areal objects of the contour map (phenomenon a) with the zones of the isoline map (phenomenon Z), $c = a \cap Z$. Indicator ρ is found for every zone starting with the one having the minimum value of level, $Z_{\min}$, and ending with the zone with its maximum value, $Z_{\max}$; $(Z_{\min}, Z_{\min} + \Delta Z, Z_{\min} + 2\Delta Z, \ldots, Z_{\min} + (n - 1)\Delta Z)$, where n is the amount of zones on the map and $n\Delta Z = Z_{\max}$. As a result we shall get a series of correlation indicators. A curve is then plotted, describing the variation of this indicator depending on the value of the zone (the values of ρ_i are plotted on the vertical axis and those of Z on the horizontal axis). If ρ is a monotonously increasing function of Z, then the interrelation is positive and if it decreases it is negative.

To obtain a more specific value of correlation between phenomena by using ρ and Z the correlation coefficient (if the correlation field is sufficiently rectilinear) or the correlation ratio is calculated. Since the zones have the same contour intervals they can be represented as a series of ordinals, i.e. $Z_1 = 1$, $Z_2 = 2, \ldots, Z_n = n$; and the correlation coefficient can then be written as

$$r = \frac{\sum_{i=1}^{n} (\rho_i - \rho_0)(Z_i - Z_0)}{n\sigma\rho \sqrt{\dfrac{(n - 1)(n + 1)}{12}}} \qquad (7.37)$$

where σ_p is the standard deviation with respect to ρ.

If the value r proves to be insufficiently representative to obtain a confidence coefficient of correlation, i.e. the evaluation of the coefficient accuracy turns out to be unsatisfactory, it is necessary to raise the number of zones by means of interpolating between the isolines.

In the automatic processing of maps by the described methods with the help of a general-purpose scanning device and a computer the elementary areas read out are formed as a result of discretization along the scanning lines in the form of raster (dotted) grids. First one map and then the other are read out. Both the maps are placed in a drum or a table in strict identity with respect to their coordinates.

7.4. Analyser of Normalized Cartographic Data

To speed up and simplify investigations with a set of maps—for example in correlation analysis—special analysers can be used. Normalized maps make it possible to read out and analyse them simultaneously (because the symbols are identified in a unidimensional space) with the help of scanners and processing blocks. The functions of the latter can be performed by specialized mini- and microcomputers.

Figure 7.9 shows the block diagram of a cartographic data analyser [130].

The analyser contains: scanner 1 designed as a rotating drum and two photoelectronic reading heads 2 and 2', synchronously connected with each other and the drum through reducer 3. The drum has special mountings 4 and 4' for maps, making it possible to displace any of the two maps relative to the reading heads by the specified values of planar coordinates Δx, Δy. When the drum is rotating, both the heads fixed on a threaded shaft move translationally along the cylinder (drum) generatrix, thereby scanning the map surface.

Signals from the outlets of reading heads, passing through amplifying and forming stages 5 and 5', enter, each via its own channel, into symbol identification blocks 6 and 6'. These are conventional devices for the identification of symbols in a unidimensional space along one scanning line; for example by colour features or raster grids.

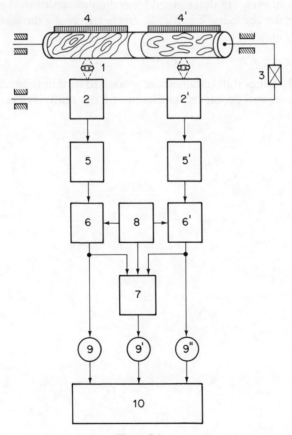

Figure 7.9

Identification signals (at the outlet of identification block), whose durations correspond to the time of the reading head passage through the identified contours, enter AND gate 7. The other entries into AND gates are connected to clock pulse generator 8. To simplify the scheme two AND gates are combined with blocks 6 and 6'. The outlets of AND gates are connected to pulse counters 9, 9' and 9", respectively, and these are connected to block 10 for the processing and recording of data.

The analyser operates in the following way.

To analyse enclosed connected areas the identification blocks are assigned with contour symbols (from the map legend) related to qualitatively specified objects or phenomena to be analysed, and when maps of isolines are analysed the zones are assigned, limited by isolines of a certain value.

As a result of the circuit operation, counter 9" records the sum total of pulses generated when reading head 2' passes over specified contours a of one map, counter 9 records the pulses related to contours b of the other map, and counter 9' those related to the areas of overlapping contours c.

The total values of contour areas, P_a, P_b, P_c for each map, as well as of the intersecting contours, are determined from the product of the sums of pulses read out from the contours, Σ_{ja}, Σ_{jb}, Σ_{jc}, respectively, by the scale factor; i.e. by elementary area P

$$P_a = P \, \Sigma_{ja}, \quad P_b = P \, \Sigma_{jb}, \quad P_c = P \, \Sigma_{jc}$$

The P_a, P_b, P_c values thus obtained are processed further in block 10 with the aim of solving various problems associated with the analysis of maps.

Conclusions

The normalization methods discussed in this monograph have been tested in experimental and operating conditions. This made possible the estimation of their economic and technical efficiency. The methods have in many respects proved to be efficient both in the preparation of cartographic data for selective input into a computer (digitization) and in map plotting at the computer output. For instance, with the method of raster discretization, ensuring a strict formalization of the raster structure of map symbols, the plotting of cartographic image on a graph plotter proved to be simpler and economically more efficient than the plotting of traditional images using arbitrary grids (rasters) in the structure of map symbols. The efficiency of digitizing newly composed normalized maps is incomparably greater than the efficiency of all the other known methods, owing to the possibility of complete automation of this process. When traditional maps are digitized the normalization methods do not always prove to be sufficiently effective either economically, because of considerably long computer times, or technically, because of insufficient reliability of symbol recognition caused by a strong influence of noises. Nevertheless, when digitizing such data as contours on topographic maps and isoanomalies on geophysical maps the normalization methods proved to be incomparably more efficient than manual digitization with the help of coders (digitizers). Normalizational methods prove to be particularly effective when the images (source data) are applied manually; for example in field and laboratory interpretation, field surveys, etc. When preparing the data for digitization in laboratory conditions (well-equipped with modern means of automation) this problem may be solved in other ways. One of the possible effective solutions is the use of the interactive processing of data, both digital and graphical. Joint use of interactive data processing and a graphical video display with scanning devices and computers makes it possible to perform with sufficient effectiveness the conversion of images into digital form and their editing for map compilation. However, even when this kind of equipment is used it is still very important to apply the methods of cartographic image normalization. In this case the normalization techniques can be simpler.

Normalization as such is of decisive importance in increasing the economic efficiency, accuracy and promptness of cartographic data processing. It should permeate the whole system of processing, starting from the preparation of primary data and ending with creating the data carriers and their storage.

Solution of the above technical and technological problems cannot be successful without the development of the methodology of cartography aimed at normalizing and formalizing the modelling processes. Of primary importance here is the development of new methods and the improvement of traditional ones for the cartographic representation of geographical data, including generalization and communication methods. Their development creates the theoretical base without which no real progress is possible in the application of modern automation and computer technology to composing and analysing maps. Their successful development, in its turn, exerts a decisive influence on the economic and technical efficiency of using cartographic information to solve serious problems of national economy. For experimental and production purposes, various types of normalized analytical and synthetic maps (geological maps, maps showing the land use, conditions of reclaimed lands, population of settlements, etc.) have been produced recently with the help of computers and automatic devices, and then printed on conventional map publication equipment. All of them are well perceived visually and some are outwardly indistinguishable from traditional maps. Thus, for example, normalized geological maps displayed at the 1976 International Cartographic Exhibition in Moscow (see Catalogue of the Exhibition, GUGK Publ., Moscow, 1976) were perceived by professional geologists as usual geological maps. Economic efficiency of composing normalized raster maps with the help of computers and automatic devices is higher than that of traditional maps, owing to the systematization and higher formalization of the methods of data representation and of the symbols used. Computer readout (selective data conversion into a digital code) is performed with sufficient promptness and reliability by means of scanning devices for the input of graphical data into a computer, and not complicated algorithms for the identification of raster signs. Scanning devices have long been produced industrially in large quantities, and in many countries. Modern scanners enable computer input with a resolution (discretization step) reaching 0.001 mm, which provides reliable recognition of raster signs with a line thickness of 0.05 mm and even less. Single-channel scanning devices have their limitations in the number of reliably identified symbols by their colour characteristics. Computer input and identification of multicoloured images can be performed by scanning systems of the Colourmation type (produced by Optronix). Furthermore, multichannel devices have also recently appeared. So today there is no problem of computerized plotting and reading of normalized maps, either technically or algorithmically. There exists only the problem of developing integrated specialized systemic software enabling efficient compilation and readout of various kinds of maps and the problem of providing the necessary technology to actualize the production processes. The technical means for the automatic readout of cartographic data (general-purpose and specialized), partially described in the present monograph, comply with present-day requirements, but when one thinks of the future, they naturally need further development. Most of the specialized technical means described for the reading of normalized maps have not been implemented at the level of a prototype or even a mock-up. Only one device, described in section 4.3.2, has been produced as an experimental batch. However, taking into account the fact that in practice most of the problems associated with the computer reading of maps are solved with the help of serially produced general-purpose scanners (in combination with computers), there is no need to manufacture devices to be used specifically with normalized maps, with some few exceptions. These are the devices for the readout of maps with a latent optical code, i.e. of the type described in section 4.3.2. Further development of the technical means for computer input, output, and processing of cartographic data (automated cartographic systems) will hardly result in any serious changes in the methods of data representation described above. Representation of cartographic data on paper will not lose its importance in the future. On the contrary, the normalization of maps and the resulting possibility of their automatic readout, as well as the fact that map duplication is very cheap, will raise the importance of maps, provided a high quality of colour image is ensured, represented in an easily comprehensible and graphic form. Normalized maps duplicated by printing will perform the functions of a universal data carrier available to a wide range of users.

Further development of the technique will be aimed at creating universal devices with a high resolution power. It is obvious that, with the increasing quality of image reproduction by printing means, new possibilities will arise of increasing the detailedness of the image, and this in its turn will require creating more accurate technical means of map-reading. To increase the accuracy of readout and the reliability of data identification it is expedient to use devices scanning with a laser beam.

Numerous organizational, technological, and scientific difficulties will have to be overcome to allow further development of this new trend in cartography. Years will probably pass before it is fully realized, understood, and then raised to a higher level in technical, scientific, and methodological respects as a result of more comprehensive research and development. The author is confident that this trend holds promise in the development of cartography—difficult but interesting.

Analysis of the modern state of the art in cartography shows that in future it will develop along two clearly manifested lines:

1. improvement of maps and atlases for a wide range of users of traditional scientific, informational, and educational maps where the main emphasis is placed on cartographic technology and communication aimed at colourful, graphic, and evocative data representation;
2. development of methods of map compilation and usage to solve scientific and technological problems, where the main requirement is to assure the accuracy, detailedness, and reliability of geographical data representation, as well as the prompt and accurate analysis of maps with the application of most modern scientific methods and means of automation and computer science.

References

Asterisks denote work in Russian.

1. Arnberger, E. *Die Kartographie als Wissenschaft und ihre Beziehungen zur Geographie und Geodäsie.* Wien: Grundsatzfragen Kartogr., 1970, pp. 1–28.
2. Arnberger, E. 'Neuere Forschungen zur Wahrnehmung von Karteninhalten. (Ein Bericht über einschlägige Forschungsergebnisse in Österreich)'. *Kartogr. Nachr.*, 1982, no. 4, S. 121–132.
*3. Aronov, V. I. *Methods of Geological Data Processing by Computers.* Moscow: Nedra Publ., 1977; 167 pp., with illus.
4. Akima Hitoshi. 'A new method of interpolation and smooth curve fitting based on local procedures'. *J. ACM*, 1970, N4.
5. Babcock, H. C. 'Automated cartography data formats and graphics: the ETL experience'. *Amer. Cartogr.*, 1978, no. 1, pp. 21–29.
*6. Berliant, A. M. 'On the essence of cartographic information'. *Izvestiya vsesoyuznogo geograficheskogo obshchestva*, 1978, vol. **110**, pp. 490–497.
7. Bertin, J. *Semiologie—Graphique.* Paris, 1967.
8. Bickmore, D. P. *Automatic Cartography and Planning.* London, 1971, p. 232.
9. Board, C. Models Geography. *Maps of Models*, London, 1967.
10. Board, C. 'Map reading tasks appropriate in experimental studies in cartographic communication'. *Can. Cartogr.*, 1978, no. 1, pp. 1–12.
11. Board, C. 'Cartographic communication'. *Cartographica*, 1981, no. 2, p. 18.
12. Boulle, F. A. 'A model of scientific data bank and its applications to geological data'. *Proceedings of the International COGEDATA Symposium*, Paris, 24–26 February 1975. University of Syracuse, New York, 13 pp.
13. Bouille, F. 'Cartographie thematique informatique—applications'. *Bol. Soc. Geogr. Lisboa*, 1978, nos. 1–3, 4–6, 5–54.
14. Bouille, F. 'Actual tools for cartography today'. *Cartographica*, 1982, no. 2, pp. 27–32.
15. Bonarius, H. 'Application of computer-assisted analysis of landsat data in semi-arid and arid areas'. *Z. Bewässerungswirt*, 1978, no. 2, pp. 131–152.
16. Boyle, A. T. 'Automatic line digitization'. *Proc. Comm. Geogr. Data Sensing and Process.* Moscow, 1976; Ottawa, 1977, pp. 53–61.
17. Boyle, A. R. 'Concerns about the present applications of computer-assisted cartography'. *Cartographica*, 1981, no. 1, pp. 31–33.
18. Boyle, R. A. 'The present status and future of scanning methods for digitization, output drafting and interactive display and edit cartographic data'. 14 Congr. Int.

249

Soc. Photogramm., Hamburg, 1980. *Int. Arch. Photogramm*, vol. 23, part 84, comm. 4. Hamburg, 1980, pp. 92–99.

*19. Boiko, A. V. *Methods and Means for the Automation of Topographic Surveys.* Moscow: Nedra Publ., 1980, 221 pp., with illus.

*20. Bocharov, M. K. *Fundamentals of the Theory of Designing Cartographic Symbol Systems.* Moscow: Nedra Publ., 1966, 135 pp., with illus.

21. Boursier, P., and Scholl, M. 'Performance analysis of compaction techniques for map representation in geographic data bases'. *Comput. Graph.*, 1982, no. 2, pp. 73–81.

22. Carstensen, L. W. 'A continuous shading scheme for two-variable mapping'. *Cartographica*, 1982, nos. 3–4, pp. 53–70.

23. Christ, F. 'Entwicklungen in der kartographischen Datenverarbeitung von 1969 bis 1979. *Kartogr. Nachr.*, 1979, no. 6, pp. 250–253.

24. Clark, R. A. 'Cartographic applications of the staran associative array processor'. *Proc. Amer. Congr. Surv. and Mapp. Fall Conv.*, Albuquerque, New Mexico, 1978. Washington, s.a., p. 95.

25. Davis, L. S. 'Representation and recognition of cartographic data'. *Map Data Process.* Proc. NATO Adv. Study. Inst., Maratea, 18–29 June 1979. New York, e.a., 1980, pp. 169–189.

26. Denegre, J. 'Different methods of image processing for assistance in interpretation'. *TRRL Suppl. Rept.*, 1982, no. 690, pp. 125–132.

*27. Derevits, V. D., Zotov, G. A., Yeremeyev, V. V., Zlobin, V. K. 'A system of digital processing of images—a means of integrated automation of map compilation process. *Izvestiya vuzov. Geodeziya i aerofotosyomka*, 1982, no. 2, p. 114.

28. Dobson, C., Ulaby, F., Stiles, I., Moor, R. K., and Holtzman, I. 'Resolution requirements for a soil moisture imaging radar'. *Int. Geosci. Remote Sens. Symp.* (IGARSS '81) Washington DC, 1981, Digest, vol. 1, pp. 427–436.

29. Dorn, M. 'Methodische Probleme beim automatischen Digitalisieren geowissenschaftlicher Karten. *Nachr. Karten- und Vermessungsw.*, 1981, B. 1, no. 85, pp. 21–29.

30. Elachi, C. 'The shuttle imaging radar (SIR-A) sensor and experiment'. *Int. Geosci. Remote Sens. Symp.* (IGARSS '82), Munich, 1–4 June 1982, Digest, vol. 2, New York, 1982, FA 6.5/1–FA 6.5/6.

31. Flanagan, G. F., Tilton, E. L. 'Landsat-D thematic mapper simulator'. *Int. Geosci. Remote Sens. Symp.* (IGARSS '82), Munich, 1–4 June 1982, Digest, vol. 1, New York, 1982, Wpl. 2/1–Wpl. 2/4.

32. Frank, A. 'Datenspeicherung für schnellen Zugriff auf Daten räumlich benachbarter Objekte'. *Nachr. Karten- und Vermessungsw.*, 1981, no. 85, pp. 37–47.

33. Gallaway, R. I. 'Das automatische Digitalisierungssystem FASTRAK'. *Nachr. Karten- und Vermessungsw.*, 1981, no. 85, pp. 49–52.

34. Giebels, M., and Weber, W. 'Höhenliniendigitalisierung nach Verfahren der Raster-Datenverarbeitung'. *Nachr. Karten- und Vermessungsw.*, 1982, B. 1, no. 88, pp. 61–75.

35. Giraund, A., Maniere, R., Monget, I. M. 'Un système informatique d'aide à recherche et à la décision applique à la gestion des milieux naturels: mese en place et fonctionnement d'une banque de données cartographiques de l'environnement (projet Molières)'. *Ecol. Mediter.*, 1983, no. 1, pp. 101–103, 112–134.

36. Goodrick, B. E. 'What is cartography?' *Cartography (Austral.)*, 1982, no. 3, pp. 146–150.

37. Goodchild, M. F., and Moy Wai-See. 'Estimation from grid data: the map as a stochastic process'. *Proc. Comm. Geogr. Data Sensing Process.* Moscow, 1976; Ottawa, 1977, pp. 67–81.

38. Grosschen, H.-W. 'Neue Wege zur "Automation" in der Kartographie'. *Int.*

250

Jahrb. Kartogr., 1981; *Int. Kartogr. Vereinig.* Bd 21. Bonn–Bad Godesberg, 1981, pp. 97–120.

39. Grygorenko, W. 'Terkepmodell kialakitasanak koncepcioja. *Foldr. kozl.*, 1980, nos. 1–2, 25–32.
40. Hajek, M., and Mitasova, I. 'Creation of thematic maps with applications of systems and computers'. Paper presented to the 10th International Conference of the ICA, 1980, Tokyo.
41. Harris, I. F., Kittler, I., Llewellyn, B., and Preston, G. 'A modular system for interpreting binary pixel representation of line-structured data on maps'. *Cartographica*, 1982, no. 2, pp. 145–175.
42. Herring, B. E., and Debinder, J. G. 'A computer-aided modeling system for geobased data'. *Comput. Ind. Eng.*, 1981, no. 3, pp. 141–152.
*43. Gokhman, V. M., and Mekler, M. M. 'On the development of the theory of language for thematic mapping'. Paper presented at VIII Int. Cartogr. Conf. Moscow, 1976; 11 pp. with illus.
*44. Gokhman, V. M., and Mekler, M. M. 'Information theory and thematic mapping'. *Voprosy geografii*, 1971, no. 88, pp. 172–183, with illus.
45. Haraliek, R. M. 'A spatial data structure for geographic information systems'. Map Data Process. *Proc. NATO Adv. Study Inst.*, Maratea, 18–29 June 1979. New York e.a., 1980, pp. 63–99.
46. Holekamp, R. A. 'Interactive computer-aided cartography without a programmer'. *Proc. Amer. Congr. Surv. Mapp. Fall Conv.*, Seattle, Wash., 1976. Falls Church, Va, 1976, pp. 363–369.
47. Heijdt, Th. van der Weiden F. van der Het. 'Grafisch interactief systeem'. *Bull. Vangroep Kartogr.*, 1977, no. 6, pp. 1/1–1/11.
48. Hoffmann, F. 'Digitale Abbildung kartographischer Datenstrukturen beim Aufbau von multivalent nutzbaren Datenstrukturen nach dem Regionalprinzip'. *Wiss. Kolog.*, 25 Jahre Hochschulhaus bild., Fachricht. kartogr., Dresden, May 1982, p. 27.
49. Kolacny, A. 'Cartographic information—a fundamental concept and term in modern cartography'. *Cartogr. J.*, 1969, no. 1, pp. 47–49.
50. Kolacny, A. 'Cartographic information—concept and term of modern cartography'. *Cartographica*, 1977, no. 19, pp. 35–45.
51. Kretshmer, I. 'Theoretical cartography: position and tasks'. *Int. Jahrb. Kartogr.*, 1980, vol. 20, pp. 142–155.
52. Krcho, J., 'Morphometric analysis of relief on the basis of geometric aspect of field theory'.
53. Krcho, J., 'Mapa ako abstraktny kartografisky model S_k geografiskej krajiny ako realneho priestoroveho systemu S_6'. *Geogr. Cas.*, 1981, no. 3, pp. 244–272.
54. Kubo, S., and Iki, K. 'ALIS: a retrieval geographic information processing and mapping system'. *24th Int. Geogr. Congr.*, Tokyo, 1980. Main Session, Abstrs, vol. 2. pp. 232–233.
*55. Kulindishev, V. A., and Malishev, Yu. F. 'An attempt at formalizing the concept of "map" (on examples of geological and geophysical maps)'. *Geologiya geofizika*, 1973, no. 6, pp. 52–59.
56. Liu, Maan-nan. 'Semi-automated mapping system (SAMS) for resource mapping and data management'. *J. Forest.*, 1982, no. 4, pp. 224–225.
57. Lybanon, M., Brown, R. M., and Gronmeyer, L. K. 'Recognition of handprinted characters for automated cartography: a progress report'. *Proc. Soc. Photo- Opt. Instrum. Eng.*, 1979, pp. 165–174.
*58. Lopatuchin, B. S., Chigirev, A. A., and Shiriayev, E. E. 'Device for read-out of isoline maps'. Authors' Cert. No. 456283, 1974. *Bull. izobr.*, no. 1, 1975.
59. Lebert, F. W., and Olson, D. 'Raster scanning for operational digitizing of graphical data'. *Photogramm. Eng. Remote Sens.*, 1982, no. 4, pp. 615–627.

60. Leberl, F., Kropatsch, W., and Lipp, V. *Int. Arch. Photogramm.* 14th Congr., Hamburg, 1980, vol. 23, part 83. Commiss. 3. Hamburg, 1980, pp. 458–468.

61. McEwen, R. B., and Calkins, H. W., 'Digital cartography in the USGS National Mapping Division: a comparison of current and future mapping processes. *Cartographica*, 1982, no. 2, pp. 11–26.

62. Mitchell, B., Guptill, S. C., Anderson, K. E., Fegeas, R. G., and Hallan, C. A. 'GIRAS: A geographic information retrieval and analysis system for handling land use and land cover data'. *Geog: Surv. Process, Pap.*, 1977, no. 1059, 16 pp. with illus.

63. Meine, K.-H. 'Kartographische Kommunikationsketten und kartographisches Alphabet'. *Mitt. Österreich. Geograph. Ges.*, 1974, no. 3, p. 116.

64. Meine, K.-H. 'Certain aspects of cartographic communication in a system of cartography as a science'. *Int. Jahrb. Kartogr.*, 1978, no. 18, pp. 102–117.

65. Meine, K.-H. 'Cartographic communication links and a cartographic alphabet'. *Cartographica*, 1977, no. 19, pp. 72–91.

66. Mor, M., and Lamdan, T. 'A new approach to automatic scanning of contour maps'. Communications of the ACM, 1972, no. 9, pp. 809–812.

67. Morrison, J. L. 'The science of cartography and its essential processes'. *Int. Jahrb. Kartogr.* Bd. 16, Bonn-Bad Godesberg, 1976, pp. 84–97.

68. Morrison, J. L. 'The science of cartography and its essential processes'. *Cartographica*, 1977, no. 19, pp. 59–71.

69. Morrison, J. L. 'Systematizing the role of "Feedback" from the map percipient to the cartographer in cartographic communication models'. Paper presented to the 10th International Conference of the ICA, 1980, Tokyo.

70. Muller, J. C. 'Bertin's theory of graphics (a challenge to North American thematic cartography'. *Cartographica*, 1981, no. 3, pp. 1–8.

71. Muehroke, P. C. 'Concepts of scaling from the map reader's point of view'. *Amer. Cartogr.*, 1976, no. 2, pp. 123–141.

72. Nagy, G., and Wagle, S. 'Geographic data processing'. *Comput. Surv.*, 1979, vol. 11, no. 2, pp. 139–181.

73. Oest, K., and Knobloch, P. 'Untersuchungen zu Arbeiten aus der Thematischen Kartographie mit Hilfe der EDV. 2 Teil. (Veröff. Akad. Raumforsch. und Landesplan. Abhandlungen, 74)'. Hannover, Hermann Schroedel Verl. KG, 1976, 411 S.

74. Otmeling, F. J. 'Einige Aspekte und Tendenzen der modernen Kartographie'. *Kartogr. Nachr.*, 1978, no. 3, pp. 90–95.

75. Ogrissek, R. 'Erkenntnistheoretische Grundlagen und Erkenntnisgewinnung in der Kartographie'. Dresden, Technical University, 1982, 78 S., illus.

76. Owczarczyk, J. 'Techniki cyfrowania map konturowych'. *Prz. geol.*, 1982, no. 3, pp. 129–131.

77. Ostrowski, W., and Ostrowski, J. A. 'Térképi információn kiválasztának fokozatai és a térkéjelek szerkesztésnek általános szabáyai'. *Folder. Kozl.*, 1980, nos. 1–2, pp. 45–49.

78. Papay, G. 'Hierarchy of representation form reflection spatial conditions'. *Hung. Cartogr. Stad.* 1980. Dedicat 10th conf. ICA. Tokyo, 1980; Budapest, 1980, pp. 29–40.

79. Pasquier, B. 'Une chaîne informatique pour la production intensive de cartes thématiques'. *Bull. Inform. Inst. Géogr. Nat.*, 1980, no. 42, pp. 25–30.

80. Penquet, D. J. 'An examination of techniques for reformatting digital cartographic data. Part 1: the raster-to-vector process'. *Cartographica*, 1981, no. 1, pp. 34–48.

81. Peterson, M. P. 'An evaluation of unclassed crossed-line choropleth mapping'. *Amer. Cartogr.*, 1979, no. 1, pp. 21–27.

252

82. Penquet, D. J. 'An examination of techniques for reformattting digital carto-graphic data: Part 2: The vector-to-raster process'. *Cartographica*, 1981, no. 3, pp. 21–33.
83. Pfrommer, W. L. 'Entwicklung, Ausgabenstellung und Ziele der Kartographi-schen Automation'. *Nachr. Karten- und Vermessungsw.*, 1977, no. 73, pp. 143–148.
84. Ratajski, L. 'The main characteristics of cartographic communication as a part of theoretical cartography. *Int. Yearb. Cartogr.*, vol. 18. Bonn, Bad Godesberg, 1978, pp. 21–32.
85. Robinson, A. H., and Petchenik, B. B. 'The map as a communication system'. *Cartographica*, 1977, no. 19, pp. 92–110.
86. Robinson, A. H. 'Cartography to the Year 2000'. *Repts. Dep. Geod. Sci.*, 1977, pp. 255–269.
87. 'Remote sensing from space—II'. *Civ. Eng. (USA)*, 1980, no. 4, pp. 63–65.
87ª. Seifert, H. 'Softwarekonzept für ein interaktives graphisches System zum Einsatz in der grossmassstabigen topographischen Kartographie'. *Nachr. Karten- und Vermessungsw.*, 1982, R. 1, no. 89, pp. 113–119.
*88. Sailshchev, K. A. 'Ideas and theoretical problems in cartography of the 80s'. *Itogi nauki i tekhniki, seriya Kartografiya*, vol. 10, 156 pp.
89. Schulz, B.-S. 'Untersuchungen und Ergebnisse zur automatischen und inter-aktiven Linienverfolgung'. *Nachr. Karten- und Vermessungsw.*, 1982, no. 87, pp. 47–62.
90. Schwenk, W. 'Kartentheoretische Grundlagen bei der Strukturierung digitaler Kartenmodelle'. 14 Congr. Int. Soc. Photogramm. *Int. Arch. Photogramm*, vol. 23, part 84, Comm. 4, Hamburg, 1980, pp. 629–639.
91. Shilin, B. V. 'Thermal infrared actial survey'. *Proc. Unat. Nat. Train. Semin. Remote Sens. Appl.*, Baku, 17–19 November 1980, Baku, 1982, pp. 146–180.
*92. Shibanov, G. P., Zhukov, V. M., and Shiriayev, Ye. Ye. 'Coding and remote acquisition of cartographic data'. *Proc. VIII Int. Conf. Moscow*, 1976, 9 pp.
*93. Shiriayev, Ye. Ye. 'Electronic engraving device'. Author's cert. No. 192664. *Bull. izobr.*, no. 5, 1967.
*94. Shiriayev, Ye. Ye. 'On the further improvement of specialized maps'. *Geodeziya i kartografiya*, 1966, no. 12, p. 51–56, with illus.
*95. Shiriayev, Ye. Ye. 'Information retrieval from gaps with the help of photoelectric and electromagnetic read-out systems'. *Izvestiya vuzov, Geodeziya i aerofotosyomka*, 1967, no. 4, pp. 65–76, with illus.
*96. Shiriayev, Ye. Ye. 'Application of luminescence on maps'. *Geodeziya i karto-grafiya*, 1967, no. 3, pp. 62–67, with illus.
*97. Shiriayev, Ye. Ye. 'Technical problems of the automation of geographical investigations on maps'. *Mathematical methods in Geography*. Proc. I All-Union Conf. on Math. Methods in Geogr. MGU Publ., 1968, pp. 150–154.
*98. Shiriayev, Ye. Ye. 'Automation of data read-out from maps by recognizing symbols by their chromatic features'. *Izv. vuzov, Geodeziya i aerofotosyomka*, 1968, no. 6, pp. 126–133, with illus.
*99. Shiriayev, Ye. Ye. 'A method for the recognition of objects on a map'. Author's Cert. No. 246921, *Bull. isobr.*, no. 21, 1969.
*100. Shiriayev, Ye. Ye. 'Photoelectronic device for the read-out of conventional signs from maps'. Author's Cert. No. 251250, *Bull. isobr.*, no. 27, 1969.
*101. Shiriayev, Ye. Ye. 'Luminescent paint'. Author's Cert. No. 254695, *Bull. isobr.*, no. 32, 1969.
*102. Shiriayev, Ye. Ye., Maslakov, I. D., and Komissarov, V. V. 'A method of obtaining coded maps'. Authors' Cert. No. 279070, *Bull. isobr.*, no. 26, 1970.
*103. Shiriayev, Ye. Ye. 'A method of obtaining luminescing maps'. Author's Cert. No. 269170, *Bull. isobr.*, no. 15, 1970.

253

*104. Shiriayev, Ye. Ye. 'The problems of "cybernizing" geographical investigations'. *Estimation of natural resources of Siberia and the Far East*. Proc. IV Conf. of Geographers of Siberia and the Far East, Novosibirsk, 1969, pp. 57–60.

*105. Shiriayev, Ye. Ye. 'Cartographic representation of discrete data (for computer and visual read-out)'. *Vestnik MGU, ser. geograf.*, 1971, no. 4, pp. 58–64, with illus.

106. Shiryaev, E. E. 'Multilevel principle of generalization and the problems of automatic map reading'. International Geography 1972. Papers submitted to the 22nd International Geographical Congress, Montreal, Canada, 1972.

*107. Shiriayev, Ye. Ye. 'A method for the automatic measurement of contour areas and perimeters with the help of scanning devices and computers'. *Geodeziya i kartografiya*, 1973, no. 5, pp. 65–72, with illus.

*108. Shiriayev, Ye. Ye. 'A method for the automatic plotting of maps showing the density of linear objects'. *Geodeziya i kartografiya*, 1973, no. 8, pp. 54–57, with illus.

*109. Shiryaev, Ye. Ye. 'Cartographic representation of data on linear objects and the principles of their automatic read-out and conversion'. *Izvestiya vuzov, Geodeziya i aerofotosyomka*, 1973, no. 5, pp. 111–120, with illus.

*110. Shiryaev, Ye. Ye., and Petrov, A. B. 'Automatic compilation of maps by the method of multiscale data representation'. *Izvestiya vuzov, Geodeziya i aerofotosyomka*, 1974, no. 4, pp. 97–104, with illus.

*111. Shiryaev, Ye. Ye. 'Normalization of the maps of isolines and their automatic reading with the help of scanning devices and computers'. *Geodeziya i kartografiya*, 1975, no. 7, pp. 74–80, with illus.

*112. Shiriayev, Ye. Ye. 'Tracing the elements of tectonics on maps, aero- and aerospace photographs with the help of computers'. *Razvedka i okhrana nedr*, 1975, no. 3, pp. 16–21, with illus.

*113. Shiriayev, Ye. Ye. 'Automation of cartographic methods for the representation and analysis of geographical data'. *Izvestiya AN SSSR, seriya geografich.*, 1975, no. 3, pp. 49–60, with illus.

*114. Shiriayev, Ye. Ye. 'Determination of volumes from maps of isolines with the help of a specialized scanning device and a computer'. *Izvestiya AN SSSR, seriya geografich.*, 1976, no. 1, pp. 118–122, with illus.

*115. Shiriayev, Ye. Ye. 'Automatic reading of data represented as limited connected regions (qualitative background)'. *Izvestiya vuzov, Geodeziya i aerofotosyomka*, 1976, no. 3, pp. 75–85, with illus.

*116. Shiriayev, Ye. Ye. 'Determination of the interrelation between phenomena by a set of maps'. *Izvestiya vuzov, Geodeziya i aerofotosyomka*, 1977, no. 5, pp. 117–122, with illus.

*117. Shiriayev, Ye. Ye. 'A possible way for computer generalization of linear objects (using a river network as an example). *Izvestiya vuzov, Geodeziya i aerofotosyomka*, 1977, no. 2, pp. 86–95, with illus.

118. Shiryaev, E. E. 'Principle of construction of ASQPSGEOGRAPHY, a single automatic system for gathering, processing and storing geographic information'. *Proceedings of the Commission on Geographical Data Sensing and Processing, Moscow, 1976*, edited by R. F. Tomlinson. A Publication of the Commission on Geographical Data Sensing and Processing of the International Geographical Union, Canada, 1977.

119. Shiryaev, E. E. 'Theory and practice of cartographic representation and analysis of information in normalized form by computers'. *International Yearbook of Cartography*, Gutersloh, 1977, vol. 17, pp. 155–164.

*120. Shiriayev, Ye. Ye. *New Methods of Geographical Data Cartographic Representation and Analysis with the Use of Computers*. Moscow: Nedra Publ., 1977, 182 pp. with illus.

*121. Shiriayev, Ye. Ye. 'Discrete data representation by the method of multiparametric continuous cartogram'. *Geodeziya i kartografiya*, 1977, no. 4, pp. 56–62, with illus.

122. Shiriayev, Ye. Ye. (Shiryaev, E. E.). 'Normalization from isoline maps and reading by means of scanner and computers'. *Geodesy, Mapping and Photogrammetry*, 1978, vol. 17, no. 1, pp. 17–20.

123. Ŝirjaev, E. E. (Shiryaev, E. E.). 'Zur Problematik automatisierter Kartographischer Systeme'. *Vermessungstechnik*, 1977, no. 10, pp. 331–335.

*124. Shiriayev, Ye. Ye. 'Normalization of field and aerospace cartographic data for automatic map plotting'. *Izvestiya vuzov, Geodeziya i aerofotosyomka*, 1979, no. 4, pp. 81–85, with illus.

125. Shiryayev, Ye. Ye. (Shiryaev, E. E.). 'A possible way for computer generalization of linear objects using a river network as an example'. *Geodesy, Mapping and Photogrammetry*, 1979, vol. 18, no. 2, pp. 99–104.

126. Shiryaev, E. E. 'Die Bestimmung von Volumen aus den Isolinien einer Karte mit Hilfe eines speziellen Abtastgerates und eines elektronischen Rechners'. *Nachrichten aus dem Karten- und Vermessungswesen*, S. I, no. 78, 1979.

127. Shiryaev, E. E. 'Problems of complete automation in the process of map making and reading by a "Closed Cycle". *International Yearbook of Cartography*, 1979, vol. 19, pp. 113–118.

*128. Shiriayev, Ye. Ye. 'Designing an optimum system of discrete symbols'. *Geodeziya i kartografiya*, 1980, no. 4, pp. 57–60, with illus.

*129. Shiriayev, Ye. Ye. 'Automation of the cartographic representation and analysis of geographical data with the use of remote sensing methods'. Papers presented at VII Conf. All-Union Astron.-Geod. Soc., Alma-Ata, 1980.

*130. Shiriayev, Ye. Ye. 'Cartographic data analyzer'. Author's Cert., *Bull. izobr.* no. 40, 1975.

*131. Shiriayev, Ye. Ye., and Vasilyev, A. S. 'Methods of normalizing traditional isoline maps in automatic construction of digital models'. *Geodeziya i kartografiya*, 1982, no. 7, pp. 34–38, with illus.

*132. Shiriayev, Ye. Ye. 'On the automation of remote sensing and cartographic data processing for water and reclamation management'. In: *Economics of Water Management*, 1982.

*133. Shiriayev, Ye. Ye. 'New aspects of the theory of cartography resulting from its cybernitization'. In: *Investigations in Geodesy, Aerial photography and Cartography*, 1982, pp. 30–38.

134. Smith, D. G. 'Raster data development in the national mapping division U.S. geological survey'. Techn. Pap. Amer. Congr. Surv. Mapp. Fall Techn. Meet., San Francisco, 9–11 September 1981; Honolulu, 14–16 September 1981. Falls Church. Va., 1981, pp. 284–288.

135. Steiner, D. 'A minicomputer-based geographical data processing system'. *Map Data Process.*, Proc. NATO Adv. Study Inst., Moratea, 18–29 June 1979. New York e.a., 1980, pp. 1–25.

136. Steward, H. J. 'Cartographic generalisation. Some concepts and explanation'. *Cartographica*, 1974, no. 10, 78 pp, with illus.

*137. Sukhikh, V. I., Sinitsin, S. G., Apostolov, Yu. S. *et al. Aerospace Methods in Nature Conservation and Forestry*. Moscow: Lesnaya promyshlennost', 1979, 286 pp., with illus.

138. Thompson, Ch. 'Digital mapping in the Ordnance Survey 1968–1978'. *Chartered Land Surv./Chartered Miner. Surv.*, 1979, no. 2, pp. 15–28.

139. Tomlinson, R. F., Calking, H. W., and Marblen, D. F. *Computer handling of Geographical Data*. Paris: UNESCO Press, 1976, p. 214.

140. Tobler, W. R. 'A transformational view of cartography'. *Repts. Dep. Geod. Sci.*, 1977, pp. 270–280.

141. Turke, K. 'Digitizing of geometric data for thematic mapping: state-of-the-art and future development'. *Nachr. Karten- und Vermessungsw.*, 1978, no. 35, pp. 85–91.

142. Unwin, D. Y. 'The computer in contemporary cartography'. *SUC Bull.*, 1981, no. 1, pp. 49–52.

*143. Vasilyev, A, S. 'Algorithm for the construction of a digital model from a normalized map of isolines'. *Izvestiya vuzov, Geodeziya i aerofotosyomka*, 1978, no. 6, pp. 91–100, with illus.

*144. Vasilyev, A, S. and Nefedov, V. V. 'Algorithm of data representation by the method of multiparametric continuous cartogram'. *Izvestiya vuzov, Geodeziya i aerofotosyomka*, 1980, no. 6, pp. 80–85, with illus.

*145. Vasilyev, A, S. 'Automation of digital modelling of the relief from traditional isoline maps'. In: *Investigations on Geodesy, Aerophotography and Cartography*.

*146. Vasilyev, A, S., and Shiriayev, Ye. Ye. 'Methods of computer interpolation of scanned isolines'. *Izvestiya vuzov, Geodeziya i aerofotosyomka*, 1983, no. 6.

*147. Vasmut, A. S. *Modelling in Cartography with the Application of Computers.* Moscow: Nedra Publ., 1983, 200 pp., with illus.

148. Weigel, I. 'Beitrag zur automatisieren Generalisierung von Wasserlaufen in allgemein-geographischen Karten. *Vermessungstechnik*, 1981, no. 8, pp. 267–269.

149. Werner, H. 'Numerical algorithms for interpolation and smoothing'. *Map Data Process.* Proc. NATO Adv. Study Inst., Maratea, 18–29 June 1979. New York e.a., 1980, pp. 331–353.

150. Weber, W. 'Rechnergestutzte Landkartenherstellung'. *Zeitschrift Vermessungsw.*, 1979, no. 1, 104 Jahrgang, pp. 1–13.

151. Weber, W. 'Optimal approximation in automated cartography'. *Optim. Estimat. Approximat. Theory.* New York–London, 1977, pp. 201–213.

152. Weber, W. 'Automationsgestutzte Generalisierung. *Nachr. Karten- und Vermess-ungsw.*, 1982, B. 1, no. 88, pp. 77–109.

153. Weber, W. 'Raster-Datenverarbeitung in der Kartographic'. *Nachr. Karten- und Vermessungsw.*, 1982, B. 1, no. 88, pp. 111–190.

154. Wehde, M. 'Grid cell size in relation to errors in maps and inventories produced by computerized map processing'. *Photogramm. Eng. Remote Sens.*, 1982, no. 8, pp. 1289–1298.

155. Williams, A. V. 'Interactive cartogram production on a microprocessor graphics system'. *Proc. Amer. Congr. Surv. Mapp. Fall Conv.*, Albuquerque, New Mexico, 1978. Washington, s.a., pp. 426–431.

*156. Vu Bik Van. 'Some new aspects of the cartographic representation of population by the method of raster discretization'. *Izvestiya vuzov, Geodeziya i aerofotosyomka*, 1977, no. 6, pp. 102–107, with illus.

*157. Vu Bik Van. 'On the accuracy of data representation by the method of raster discretization'. In: *Investigations on Geodesy, Aerophotography and Cartography*, 1082, pp. 79–84, with illus.

158. Yoeli, P. 'Digital terrain models and their cartographic and cartometric utilization'. *Cartogr. J.*, 1983, no. 1, pp. 17–22.

Index